MERRILL

CHEMISTRY

STUDY GUIDE
STUDENT EDITION

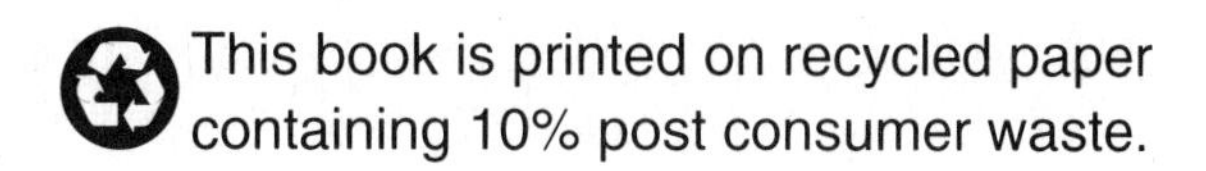

GLENCOE

Macmillan/McGraw-Hill

New York, New York   Columbus, Ohio   Mission Hills, California   Peoria, Illinois

A GLENCOE PROGRAM

- Student Edition
- Teacher Wraparound Edition
- Laboratory Manual–Student Edition
- Laboratory Manual–Teacher Edition
- English/Spanish Glossary
- Computer Test Bank
- Videodisc Correlations
- Transparency Package
- Problems and Solutions Manual
- Solving Problems in Chemistry
- Study Guide–Student Edition
- Mastering Concepts in Chemistry–Software

Teacher Resource Books:
- Enrichment
- Critical Thinking/Problem Solving
- Study Guide–Teacher Edition
- Lesson Plans
- Chemistry and Industry
- ChemActivities
- Vocabulary and Concept Review
- Transparency Masters
- Reteaching
- Applying Scientific Methods in Chemistry
- Evaluation

Send all inquiries to:

GLENCOE DIVISION
Macmillan/McGraw-Hill
936 Eastwind Drive
Westerville, OH 43081

ISBN 0-02-827226-9

Printed in the United States of America.
1 2 3 4 5 6 7 8 9 0 - MAL - 02 01 00 99 98 97 96 95 94

CONTENTS

TO THE STUDENT

This ***Study Guide*** for *Merrill Chemistry* provides you with an easy way to learn chemistry. It includes lessons for each section of your textbook. Each worksheet closely follows the chapter text and helps you understand the new vocabulary and major concepts presented.

You will find the directions simple and easy to follow. The diagrams are clear and useful. You may be asked to label a diagram, complete a table, fill in the blanks, match, or answer true or false to important information contained in your subject matter. Complete each Study Guide section after you have read the chapter section assignment.

Name ______ Date ______ Class ______

Chapter 1

STUDY GUIDE

THE ENTERPRISE OF CHEMISTRY

Complete each sentence.

1. Matter is anything that has the property of ______.
2. The way that matter behaves is described by its ______.
3. The property possessed by all matter, which in the proper circumstances can be made to do work, is called ______.
4. A property of matter that shows itself as a resistance to any change in motion is called ______.
5. Potential energy depends upon the ______ of an object with respect to some reference point.
6. Kinetic energy is the energy possessed by an object because of its ______.
7. Energy that is transferred as electromagnetic waves is called ______ energy.

Answer each question.

8. What do chemists do?
9. What are the two ways energy can be transferred between objects?
10. What does the law of conservation of mass state?
11. What does the law of conservation of energy state?
12. What does the law of conservation of mass-energy state?

13. When are changes of energy to mass and mass to energy observable?

14. What is an intermediate?

15. Why would water used in the making of a consumer product be considered a raw material and not an intermediate?

16. What is a model?

17. What law does Einstein's equation $E = mc^2$ express?

18. Explain why pesticides did not succeed in permanently controlling insect populations.

19. State two harmful effects of pesticides.

Chapter 2

STUDY GUIDE

2.1 DECISION MAKING

Answer each of the following.

1. What are the two requirements that must be met before the answer to a chemistry problem can be called correct?

2. In solving problems, what should you do if the problem includes irrelevant information?

3. Why should you carefully consider the units in which your answer will be expressed?

4. Why is it important to examine your calculated answer to a problem?

5. In the table below, list the seven problem-solving steps outlined in your text. Then, read the problem and describe how to apply each of the seven steps to solve the problem.

   Problem: Seven students are working to put together a student newsletter. There are six sheets of paper in each newsletter. Three students can each assemble two newsletters per minute. Three other students can each assemble three newsletters per minute. The remaining student can assemble four newsletters per minute. How many newsletters will the group assemble in one hour?

Steps	Application of steps
(1)	
(2)	
(3)	
(4)	
(5)	
(6)	
(7)	

Name ______________________ Date ______________________ Class ______________________

Chapter 2

STUDY GUIDE

2.2 NUMERICAL PROBLEM SOLVING

1. Distinguish between quantitative and qualitative descriptions.

____

2. Complete the table.

Quantity	Unit	Unit symbol
Electric current		
	meter	
		s
Mass		
	kelvin	
Amount of substance		
Luminous intensity		cd

3. List the following units from largest to smallest: meter, millimeter, kilometer, centimeter, picometer.

____

Match each quantity to be measured with the most appropriate SI unit.

a. centimeter
b. gram
c. kelvin
d. kilogram
e. meter
f. newton
g. second

____ 4. length of a swimming pool
____ 5. length of a pencil
____ 6. mass of a pencil
____ 7. mass of a bag of apples
____ 8. time between two heartbeats
____ 9. temperature of boiling water
____ 10. weight of a textbook

11. Each of five students used the same ruler to measure the length of the same pencil. These data resulted: 15.33 cm, 15.34 cm, 15.33 cm, 15.33 cm, 15.34 cm. The actual length of the pencil was 15.55 cm. Using these numbers as examples, distinguish between accuracy and precision.

____

____

____

12. For each group of digits indicated in the three numbers below, state whether the digits are significant. Then state the rule that applies.

a b c d e f
4500.60 0.000 799 220

a. ______________________________

b. ______________________________

c. ______________________________

d. ______________________________

e. ______________________________

f. ______________________________

13. A stack of books contains 10 books, each of which is determined, by a ruler graduated in centimeters, to be 25.0 cm long. How do these two quantities differ in terms of significant digits?

____

____

____

14. What is the percent error in a determination that yields a value of 8.38 g/cm³ as the density of copper? The literature value for this quantity is 8.92 g/cm³.

15. A student measures the melting point of ammonium acetate, $NH_4C_2H_3O_2$, as 117°C, but the literature value is 114°C. What is the percent error in the measurement?

Derive units that would be appropriate for each of the following situations.

16. A faucet trickles, even when it is shut off. How would you express the rate of water flow?

____

17. When you add ice cubes to a glass of water, the water's temperature decreases. How would you express the change in temperature caused by the addition of each ice cube?

____

18. Complete the following table of densities, using Table 2.3 in your text to determine the probable identity of *X*, *Y*, and *Z*.

Material	**Mass (g)**	**Volume (cm³)**	**Density (g/cm³)**	**Identity**
X	16.0	2.0		
Y	17.8	2.0		
Z	2.7	3.4		

Chapter 3

STUDY GUIDE

3.1 Classification of Matter

Answer each of the following.

1. How does a material differ from a mixture?

2. Describe an iceberg afloat in an ocean, using the terms *phase, state,* and *system.*

3. Compare and contrast heterogeneous and homogeneous mixtures and give two examples of each.

4. Suppose a glass contains 2 cups of sugar dissolved in 4 cups of water. Classify this mixture. Which component is the solute? Which is the solvent?

5. Explain why sea water is classified as a solution.

6. What is a substance? How does it differ from a homogeneous mixture?

7. How are elements and compounds related?

8. Are compounds more or less ordered than heterogeneous mixtures?

9. Classify the following materials as heterogeneous mixtures, compounds, elements, or solutions: sugar, salt, water, gold, brass, wood, carbon, and air.

10. Classify the following substances as organic or inorganic: methane (CH_4), aluminum chloride ($AlCl_3$), ethanol (C_2H_5OH), benzene (C_6H_6), and copper (Cu).

Name ______________ Date ______________ Class ______________

Chapter 3

STUDY GUIDE

3.2 CHANGES IN PROPERTIES

Write T for true and F for false. If a statement is false, change the underlined word or phrase and write your correction on the blank.

__________ 1. A change in which the same substance remains after the change is called a chemical change.

__________ 2. Melting is a physical change.

__________ 3. Evaporating water from a saltwater solution is a chemical change.

__________ 4. A gaseous substance that forms from a solution is called a precipitate.

__________ 5. The reaction of iron with hydrochloric acid is a chemical change.

Use the figure to answer the following questions.

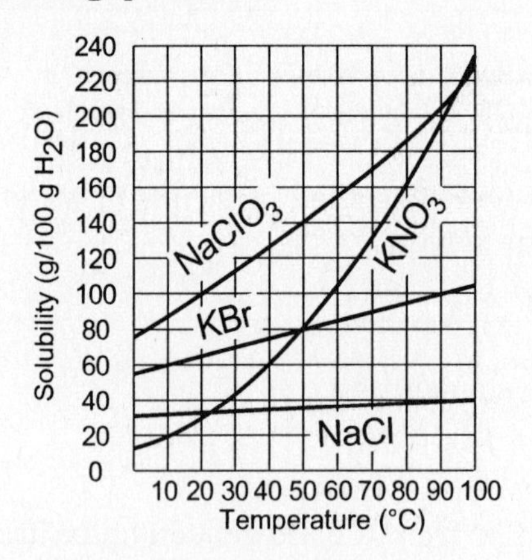

6. What is the solubility of sodium chloride at 30°C? __________

7. Which is more soluble at 90°C: KNO_3 or $NaClO_3$? __________

8. At what temperature are the solubilities of KNO_3 and KBr the same? __________

9. At what temperature will 200 g of $NaClO_3$ dissolve in 100 g of H_2O? __________

Identify each property below as either a chemical property or a physical property. For each physical property, state whether it is extensive or intensive.

10. color __________

11. width __________

12. melting point __________

13. density __________

14. resistance to an acid __________

15. conductivity __________

16. luster __________

17. flammability __________

18. weight __________

Chapter 3

STUDY GUIDE

3.3 ENERGY

Answer the following questions.

1. Recall that ordinary salt is a compound of sodium and chlorine. List several systems that can be defined in a shaker full of salt.

______

______

2. What is heat?

______

3. What is a joule?

______

4. How is the calorie related to the joule?

______

5. Contrast endothermic and exothermic reactions.

______

6. What is specific heat?

______

Solve the following problems.

7. How many joules are in 1.11 Calories?

8. An average baked potato contains 164 Calories. Calculate the energy value of the potato in joules.

9. How much heat is lost when a 4110-g metal bar whose specific heat is 0.2311 J/g·C° cools from 100.0°C to 20.0°C?

10. What is the specific heat of copper if a 105-g sample absorbs 15 200 J and the change in temperature is 377°C?

11. What is the final temperature of a system of iron and water if a piece of iron with a mass of 25.0 grams at a temperature of 75.0°C is dropped into an insulated container of water at 35.5°C? The mass of the water is 150.0 grams. The specific heat of iron is 0.449 J/g·C°, and the specific heat of water is 4.184 J/g·C°.

Chapter 4

STUDY GUIDE

4.1 *Early Atomic Models*

Complete the following sentences.

1. The ________________ is an apparatus that helped scientists discover that atoms contain electrons.

2. Isotopes of an element have the same number of protons but different numbers of ________________.

3. The number of protons in the nucleus of an atom is called the ________________ of that element.

4. The total number of protons and neutrons in a nucleus is called the ________________.

5. The spontaneous production of rays of particles and energy by unstable nuclei is called ________________.

Match the names of the scientists to their discoveries.

a. Einstein
b. Proust
c. Becquerel
d. Marie and Pierre Curie
e. Geiger, Marsden, and Rutherford
f. Millikan
g. Thomson
h. Lavoisier
i. Dalton
j. Chadwick

____ 6. ability of radium to give off rays

____ 7. positively charged nucleus

____ 8. $E = mc^2$

____ 9. electron's charge-to-mass ratio

____ 10. electron's charge

____ 11. neutrons

____ 12. law of conservation of mass

____ 13. ability of uranium to expose photographic film

____ 14. law of multiple proportions

____ 15. law of definite proportions

16. Briefly state the following:

a. law of definite proportions

____

____

b. law of multiple proportions

____

____

____

17. Complete the following table of isotopes of oxygen.

Isotope	Protons	Neutrons	Mass number
oxygen-16	8		16
oxygen-17	8	9	
oxygen-18	8		18

18. What is wrong with the diagram of a cathode-ray tube? How should it look? Explain your answer.

____

____

____

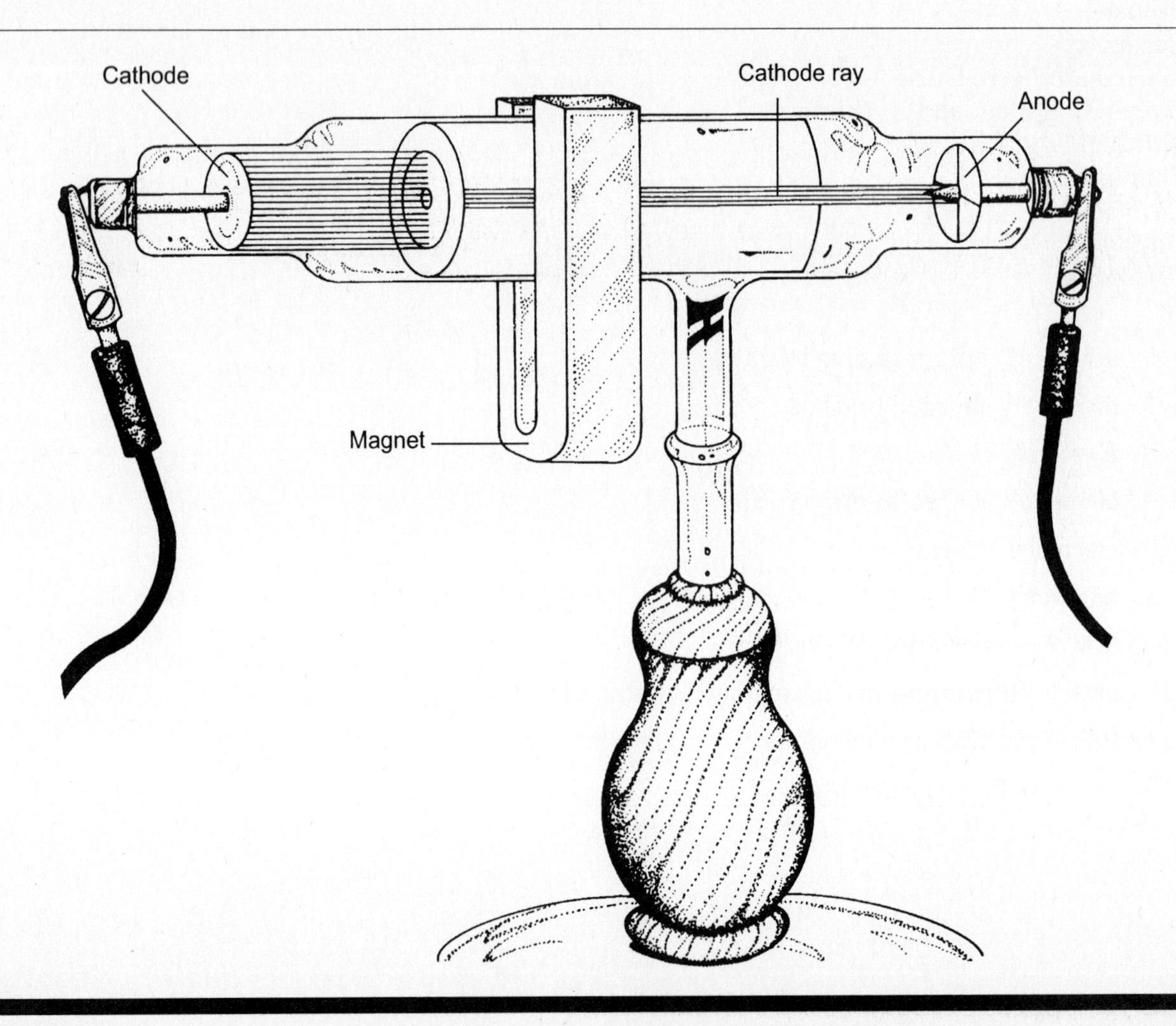

Chapter 4

STUDY GUIDE

4.2 PARTS OF THE ATOM

Write T for true or F for false. If a statement is false, replace the underlined word or phrase with one that will make the sentence true, and write your correction on the blank provided.

______ 1. Alpha radiation and beta radiation are made up of particles.

______ 2. Gamma radiation is of very low energy.

______ 3. Waves that are slightly higher in frequency than visible light are called ultraviolet rays.

______ 4. Wavelength is represented by the Greek letter nu (ν).

______ 5. Electromagnetic energy travels at the speed of sound.

______ 6. Electromagnetic waves with low frequencies have long wavelengths.

______ 7. Absorption spectra are produced by energy given off by excited gaseous atoms.

______ 8. The Rutherford-Bohr model of the atom is sometimes called the planetary model.

______ 9. Planck proposed the equation $E = mc^2$.

______ 10. A photon is a quantum of radiant energy.

Match the particles to the descriptions or examples.

a. beta particle
b. photon
c. alpha particle
d. baryon
e. meson
f. lepton
g. quarks
h. nucleon
i. antiparticle
j. gluon

____ 11. particle exchanged by quarks

____ 12. helium nucleus

____ 13. an electron, neutrino, muon, or pion

____ 14. always contains a quark and an antiquark

____ 15. a positron, for example

____ 16. an electron released by a nucleus

____ 17. always contains three quarks

____ 18. energy packet of light

____ 19. any component of a nucleus

____ 20. "up," "down," "charm," "strange," "top," or "bottom"

Complete the following sentences.

21. ______________________ includes visible light, radio waves, and infrared, ultraviolet, and X rays.

22. Alpha, beta, and gamma rays given off by nuclei are all forms of ______________________ .

23. Scientists use a nuclide of the element ______________________ as the standard for the atomic mass scale.

24. The symbol for an atomic mass unit is ______________________ .

25. The ______________________ atomic mass takes into account the differing masses and percent occurrences of isotopes of an element.

26. Inside a mass spectrometer, the paths of heavy particles are bent ______________________ than are the paths of lighter particles.

27. Calculate the average atomic mass of lithium, which occurs as two isotopes that have the following atomic masses and abundances in nature: 6.017 u, 7.30%; and 7.018 u, 92.70%.

Chapter 5

STUDY GUIDE

5.1 MODERN ATOMIC STRUCTURE

Complete the sentence or answer the question.

1. Define or explain the following.
   a. wave-particle duality of nature ______
   b. Newtonian mechanics ______
   c. quantum mechanics ______
   d. the Heisenberg uncertainty principle ______
   e. quantum numbers ______

2. The product of mass and velocity of an object is called the ______.

3. The French scientist whose hypothesis on the wave nature of particles helped lead to the present-day theory of atomic structure was ______.

4. The region in which an electron travels around the nucleus of an atom is called a(n) ______.

5. Complete the table, using de Broglie's equation, $\lambda = \frac{h}{mv}$, for the wavelength λ of a particle of mass m and velocity v. Note: Planck's constant $h = 6.63 \times 10^{-34}$ kg · m²/s.

m (kg)	v (m/s)	λ (m)
0.005 00	9.10×10^{6}	
0.0220		1.51×10^{-34}
	5.10	1.13×10^{-5}

6. Explain what is meant by the *probability* of finding an electron at a point in space.

______

Write T *for true or* F *for false. If a statement is false, replace the underlined word or phrase with one that will make the statement true, and write your correction on the blank provided.*

______________ 7. De Broglie used Einstein's relationship between matter and energy and Schrödinger's wave equation to develop his equation for wavelength of a particle.

______________ 8. To be able to give a full description of an electron, you would need to know two things: its present position and its radius.

______________ 9. It is impossible to know both the exact position and the exact momentum of an object at the same time.

______________ 10. One of the most important developments of Schrödinger's wave equation was the idea of quantum numbers, which describe the behavior of electrons.

______________ 11. In Schrödinger's wave equation, n can have only negative whole number values.

Match each mathematical expression with the correct description.

a. de Broglie's equation for the wavelength of a particle
b. Einstein's equation relating matter and energy
c. Planck's quantum equation
d. term for total energy in Schrödinger's wave equation

___ 12. $\lambda = \frac{h}{mv}$

___ 13. $\frac{2\pi^2 me^4}{h^2 n^2}$

___ 14. $E = h\nu$

___ 15. $E = mc^2$

Chapter 5

STUDY GUIDE

5.2 QUANTUM THEORY

Complete the sentence or answer the question.

1. When an electron in a hydrogen atom moves from a higher to a lower energy state, the energy difference is emitted as a quantum of _______________.

2. Define the four quantum numbers *n*, *l*, *m*, and *s*, explain what information is given by each, and describe the range of values each may take.

3. Orbitals of the same energy are said to be _______________.

4. The space occupied by one pair of electrons is called a(n) _______________.

5. What is the formula for calculating the maximum number of electrons that can occupy any energy level in an atom? _______________

6. Complete the following table.

Energy level	Number of sublevels	Number of orbitals	Maximum number of electrons
1			
2			
3			
4			

7. State the *Pauli exclusion principle.*

Name ______ Date ______ Class ______

8. Complete the following table.

Sublevel	Number of orbitals	Maximum number of electrons
s		
p		
d		
f		

9. Complete the electron configurations for the following atoms by drawing in the arrows indicating the electrons with appropriate spins for each orbital.

	1*s*	2*s*	2*p*
a. beryllium (atomic number 4)	◯	◯	
b. carbon (atomic number 6)	◯	◯	◯◯◯
c. fluorine (atomic number 9)	◯	◯	◯◯◯

Write T *for true or* F *for false. If a statement is false, replace the underlined word with one that will make the statement true, and write your correction on the blank provided.*

______ 10. If two electrons occupy the same orbital, they must have opposite spins.

______ 11. The principal quantum number describes the energy level of an electron in an atom.

______ 12. The Schrödinger wave equation is solvable for any multielectron system.

______ 13. Four quantum numbers are required to describe completely an electron in an atom.

______ 14. The sum of all electron clouds in any sublevel (or energy level) is a tetrahedral cloud.

Chapter 5

STUDY GUIDE

5.3 *DISTRIBUTING ELECTRONS*

Complete the sentence or answer the question.

1. In using the "rule of thumb" arrow diagram to predict electron arrangements for most atoms in the ground state, what assumption must be made?

 ____

 ____

2. It is important to know how to find the electron arrangement of an atom in order to predict its ______________________ .

3. The number of electrons used in writing the electron configuration of an atom of an element is equal to the element's ______________________ .

4. Write the electron configurations for the following elements:

 a. lithium ______________________

 b. boron ______________________

 c. sodium ______________________

 d. sulfur ______________________

 e. calcium ______________________

5. The notation in which outer-energy-level electrons are indicated around the symbol of an element is referred to as a(n) ______________________ .

6. Write the three steps for drawing electron dot diagrams.

 ____

 ____

 ____

 ____

 ____

7. Write the electron dot diagrams for the following elements:

 a. lithium

 b. boron

 c. sodium

 d. sulfur

 e. calcium

Name ______________________ Date ______________________ Class ______________________

Chapter 6

STUDY GUIDE

6.1 DEVELOPING THE PERIODIC TABLE

Fill in the blanks with appropriate terms.

1. The table below shows the way a scientist named ______________________ classified the elements into groups he called ______________________.

Symbol	Atomic mass	Symbol	Atomic mass
Ca	40.1	Cl	35.5
Ba	137.3	I	126.9
Ca-Ba Average	88.7	Cl-I Average	81.2
Sr	87.6	Br	79.9

2. The table below shows the classification of elements according to a principle called the ______________________, which was developed by ______________________.

1	2	3	4	5	6	7
Li	Be	B	C	N	O	F
Na	Mg	Al	Si	P	S	Cl

3. The elements in Mendeleev's table were arranged in order of increasing ______________________.

4. As a result of Henry Moseley's work, the modern periodic table is arranged according to increasing ______________________.

5. The atomic number of an element indicates the number of ______________________ in the nucleus.

Write T *for true and* F *for false. If a statement is false, replace the underlined word with one that makes the statement true.*

______________________ 6. The electron configurations of hydrogen and helium are <u>not similar</u>, so each element is in a separate column of the periodic table.

______________________ 7. The elements in columns 3 through 12 (IIIB through IIB) are called the <u>noble gases</u>.

______________________ 8. Each time a new principal energy level is started, a new <u>row</u> in the periodic table begins.

______________________ 9. The lanthanoid series contains the elements lanthanum through <u>ytterbium</u>.

______________________ 10. In the periodic table, a horizontal row of elements is called a <u>period</u>.

Answer the following questions.

11. What is the pattern of placing electrons in energy sublevels for elements in the actinoid series?

12. Why are neon and helium placed in the same column in the periodic table?

13. Compare the ways Dobereiner and Newlands classified the elements.

14. Fill in the following table.

Sublevel type	Electron capacity
	2
p	
	10
f	14

ame ____________ Date ____________ Class ____________

Chapter 6

STUDY GUIDE

6.2 USING THE PERIODIC TABLE

Match each of the following elements with the element in the list that has the most similar chemical properties. Use the periodic table as a guide.

a. zinc (Zn)
b. helium (He)
c. potassium (K)
d. rhodium (Rh)
e. terbium (Tb)
f. vanadium (V)
g. phosphorus (P)
h. gold (Au)
i. indium (In)
j. chlorine (Cl)
k. rhenium (Re)
l. oxygen (O)
m. yttrium (Y)
n. silicon (Si)
o. radium (Ra)

____ 1. cobalt (Co)

____ 2. carbon (C)

____ 3. argon (Ar)

____ 4. cerium (Ce)

____ 5. lithium (Li)

____ 6. aluminum (Al)

____ 7. iodine (I)

____ 8. copper (Cu)

____ 9. manganese (Mn)

____ 10. niobium (Nb)

____ 11. calcium (Ca)

____ 12. scandium (Sc)

____ 13. mercury (Hg)

____ 14. selenium (Se)

____ 15. arsenic (As)

	1 1A	2 IIA	3 IIIB	4 IVB	5 VB	6 VIB	7 VIIB	8	9 VIIIB	10	11 IB	12 IIB	13 IIIA	14 IVA	15 VA	16 VIA	17 VIIA	18 VIIIA
1	1 H																	2 He
2	3 Li	4 Be											5 B	6 C	7 N	8 O	9 F	10 Ne
3	11 Na	12 Mg											13 Al	14 Si	15 P	16 S	17 Cl	18 Ar
4	19 K	20 Ca	21 Sc	22 Ti	23 V	24 Cr	25 Mn	26 Fe	27 Co	28 Ni	29 Cu	30 Zn	31 Ga	32 Ge	33 As	34 Se	35 Br	36 Kr
5	37 Rb	38 Sr	39 Y	40 Zr	41 Nb	42 Mo	43 Tc	44 Ru	45 Rh	46 Pd	47 Ag	48 Cd	49 In	50 Sn	51 Sb	52 Te	53 I	54 Xe
6	55 Cs	56 Ba	71 Lu	72 Hf	73 Ta	74 W	75 Re	76 Os	77 Ir	78 Pt	79 Au	80 Hg	81 Tl	82 Pb	83 Bi	84 Po	85 At	86 Rn
7	87 Fr	88 Ra	103 Lr	104 Unq	105 Unp	106 Unh	107 Uns	108 Uno	109 Une									

LANTHANOID SERIES	57 La	58 Ce	59 Pr	60 Nd	61 Pm	62 Sm	63 Eu	64 Gd	65 Tb	66 Dy	67 Ho	68 Er	69 Tm	70 Yb
ACTINOID SERIES	89 Ac	90 Th	91 Pa	92 U	93 Np	94 Pu	95 Am	96 Cm	97 Bk	98 Cf	99 Es	100 Fm	101 Md	102 No

Write a T *for true or* F *for false. If the statement is false, replace the underlined word with a word that makes the statement true.*

______ 16. The electron configurations of all elements in Group 1 (IA) end in <u>s^1</u>.

______ 17. In the electron configuration for the outer energy level of potassium, $4s^1$, the coefficient *4* indicates the <u>group number</u>.

______ 18. Atoms with full outer energy levels are <u>very reactive</u>.

______ 19. Elements such as silicon are called metalloids because they have properties of <u>both metals and nonmetals</u>.

______ 20. Elements with three or fewer electrons in the outer energy level are usually <u>metals</u>.

Answer each of the following.

21. Describe the properties of nonmetals, and give examples.

______

______

______

22. What is the main reason that atoms react with each other?

______

______

______

23. What determines an atom's chemical properties?

______

______

24. State the octet rule.

______

______

25. Describe the general positioning of metals and nonmetals in the periodic table.

______

______

ame ______________________ Date ______________ Class ______________

Chapter 7

STUDY GUIDE

7.1 SYMBOLS AND FORMULAS

Answer each of the following.

1. What is the most common source for an element's name? What are three other possible sources for the name of an element?

____

____

2. Complete the table by filling in the missing name or chemical symbol of each element shown.

Element	Symbol
	K
gold	
phosphorus	

3. a. What element does the symbol $^{40}_{20}Ca$ represent? ______________

b. What is this element's mass number? ______________

c. How many protons, electrons, and neutrons does the element contain? ______________

d. What would the charge be if an atom of this element lost two electrons? ______________

4. Complete the table by filling in the missing information.

Chemical formula	Names of elements in compound	Relative number of each atom
NH_3		
$C_{12}H_{22}O_{11}$		
ZnF_2		
Fe_2O_3		

Write T *for true or* F *for false. If a statement is false, change the underlined word(s) to make the statement correct by writing the correct word(s) on the blank.*

______________ 5. Atoms are electrically charged particles.

______________ 6. In the formula of an ionic compound, the negative ion is written first.

______________ 7. The charge on a diatomic molecule is called its oxidation number.

8. Write the formulas of the compounds that will be formed when the positive ions listed in the table below combine with each negative ion listed.

	NO_3^-	S^{2-}	OH^-	Cl^-
Al^{3+}				
NH_4^+				
Pb^{4+}				

Name ______________________ Date ______________ Class ______________

Chapter 7

STUDY GUIDE

7.2 NOMENCLATURE

Match the terms listed with their descriptions. Write the correct letters on the lines provided.

___ 1. binary compounds
___ 2. hydrocarbons
___ 3. sulfur hexafluoride
___ 4. aluminum arsenate
___ 5. baking soda
___ 6. *meth-*
___ 7. copper(II) oxide
___ 8. common acids
___ 9. cyclopentane
___ 10. nitrogen

a. example of a binary compound formed from a metallic element with variable oxidation states
b. indicates one carbon atom
c. organic compounds composed solely of H and C
d. forms five different binary compounds with oxygen
e. common name of sodium hydrogen carbonate
f. has five carbon atoms linked in a ring
g. compounds that contain only two elements
h. common name of a compound containing six fluorine atoms
i. have names that do not normally follow the rules for naming compounds
j. demonstrates the rule that the name of a polyatomic ion does not end in *-ide*

11. Complete the table by filling in the missing names.

Compound formula	Name
KCl	
$ZnCO_3$	
C_4H_{10}	
$FeCl_3$	
N_2O_3	

Complete the sentence or answer the question.

12. What is the difference between a molecular formula and an empirical formula?

____

____

13. The formulas for ionic compounds are almost all ______________ formulas.

14. The molecular formula of a compound is always a whole-number multiple of the ______________ formula.

15. What is the empirical formula of the compound N_2O_4?

____

16. How would you use symbols to designate three formula units of zinc sulfide?

____

Chapter 8

STUDY GUIDE

8.1 FACTOR-LABEL METHOD

Complete the sentence or answer the question.

1. Define the following.
   a. factor-label method ______________________
   ____
   b. scientific notation ______________________
   ____
2. A conversion factor is a ratio equivalent to ______________.
3. What advantage does scientific notation provide?
   ____
   ____
4. List the rules for handling decimal places or significant digits in the following kinds of calculations.
   a. addition and subtraction
   ____
   ____
   b. multiplication and division
   ____
   ____
5. Complete the following conversion-factor table.

Unit given	Unit desired	Conversion factor
m	cm	
kg	g	
nm	m	
h	min	
cm^3	dm^3	
mm	m	
m^3	cm^3	

6. Convert each of the following numbers to scientific notation.

a. 124 ______ d. 0.1001 ______

b. 0.000 02 ______ e. 2.0 ______

c. 564.45 ______ f. 301.03 ______

7. Enter the number of significant digits for each of the numbers below.

a. 2.234 ______ d. 50 000 ______

b. 124.3 ______ e. 3.141 592 654 ______

c. 0.000 430 ______ f. 78 456.1010 ______

8. Using the factor-label method, perform the following conversions. Express your answers in the correct number of significant digits.

a. 30.0 min to h

b. 125.1 km/s to m/min

c. 0.003 m to cm

d. 55.0 km/h to cm/s

9. Perform the following calculations. Express your answers to the correct number of significant digits and in scientific notation.

a. add: $2.01 \times 10^{3} + 4.23 \times 10^{-1}$

______

c. multiply: $10.1 \times 10^{2} \times 6.23 \times 10^{23}$

______

b. subtract: $7.2 \times 10^{2} - 7.1 \times 10^{3}$

______

d. divide $\dfrac{5 \times 10^{-3}}{9.0 \times 10^{6}}$

______

Chapter 8

STUDY GUIDE

8.2 FORMULA-BASED PROBLEMS

Match the term with the correct description.

a. molecular mass
b. Avogadro constant
c. molarity
d. empirical formula
e. formula mass
f. molar mass
g. percentage composition
h. molecular formula

____ 1. the sum of the atomic masses of all atoms in the formula unit of an ionic compound

____ 2. the simplest ratio of the elements in a compound

____ 3. the sum of all the atomic masses in a molecule

____ 4. the mass of 6.02×10^{23} molecules, atoms, ions, or formula units of a species

____ 5. the ratio between the moles of dissolved substance and the volume of solution in cubic decimeters

____ 6. shows the actual number of atoms in a molecule

____ 7. 6.02×10^{23}

____ 8. a statement of the relative mass each element contributes to the mass of a compound as a whole

Find the formula mass or molecular mass of the following compounds.

9. ammonia, NH_3

10. methane, CH_4

11. sodium hydrogen carbonate (baking soda), $NaHCO_3$

Using the factor-label method, determine how many moles are represented by 25.0 g of each of the compounds listed in questions 9 through 11.

12. ammonia, NH_3

13. methane, CH_4

14. sodium hydrogen carbonate, $NaHCO_3$

Using the factor-label method, determine the mass, in grams, of 2.50×10^{24} molecules or formula units of each of the compounds in questions 9 through 11.

15. ammonia, NH_3

16. methane, CH_4

17. sodium hydrogen carbonate, $NaHCO_3$

18. Complete the following table by filling in the correct values.

Concentration	Moles of solute	Volume of solution
10.0M		3.00 dm^3
0.062M	0.651 mol	
	6.3 mol	5.9 dm^3
6.0M		1.0 dm^3
	0.0555 mol	0.500 dm^3

Describe the preparation of each of the following solutions.

19. 1.00 dm^3 of 6.00M NaCl

20. 2.50 dm^3 of 0.100M $BaCl_2$

Find the percentage composition of the following compounds.

21. carbon dioxide, CO_2

22. sulfuric acid, H_2SO_4

23. hydrogen peroxide, H_2O_2

Calculate the empirical formulas of the following.

24. a compound that is 45.9% K, 16.5% N, and 37.6% O

25. a compound, 5.00 g of which contains 4.28 g C and 0.720 g H

26. a compound, a 100.0-g sample of which contains 11.2 g H and 88.8 g O

Calculate the molecular formulas of the compounds listed below, given the empirical formula and formula mass of each.

27. empirical formula = CH, formula mass = 78.1 u

28. empirical formula = NH_2, formula mass = 32.1 u

Chapter 9

STUDY GUIDE

9.1 CHEMICAL EQUATIONS

Complete each sentence.

1. The process by which one or more substances are changed into one or more different substances is called a(n) ______________.

2. The starting substances in a chemical reaction are called ______________.

3. The symbol written after a formula to indicate a solid is ______________.

4. In a balanced chemical equation, the number of ______________ of any given kind on the left side is equal to the number on the right side.

Match each reaction type with the correct description.

a. combustion
b. double displacement
c. synthesis
d. single displacement
e. decomposition

____ 5. The positive and negative portions of two compounds are interchanged.

____ 6. A compound burns, reacting with oxygen.

____ 7. Two or more substances combine to form one new substance.

____ 8. A substance breaks down to form simpler substances when energy is supplied.

____ 9. One element replaces another in a compound.

Write T *for true or* F *for false. If a statement is false, replace the underlined word or phrase with one that will make the statement true, and write your correction on the blank provided.*

______________ 10. The symbol (aq) after a formula indicates that the substance is <u>dissolved in water</u>.

______________ 11. A number written to the left of a formula in an equation is called a <u>subscript</u>.

______________ 12. <u>Carbon</u> and water are typical products in a combustion reaction.

______________ 13. The substances generally written on the right in chemical equations are called <u>products</u>.

______________ 14. In a chemical equation, the symbol that stands for *yields* is <u>a plus sign</u>.

______________ 15. The general form of a <u>double displacement</u> reaction is: element + compound → element + compound.

______________ 16. In balancing a chemical equation, it <u>is</u> permissible to change subscripts.

Name ______________________ Date ______________________ Class ______________________

Chapter 9

STUDY GUIDE

9.2 STOICHIOMETRY

Complete each sentence.

1. The branch of chemistry that deals with the amounts of substances involved in chemical reactions is called ______________________.

2. In solving a mass-mass problem, one must convert the number of grams of the given substance to ______________________.

3. The actual amount of product divided by the theoretical amount and multiplied by 100 is called the ______________________.

4. A(n) ______________________ problem is one in which the amount of heat absorbed or released during a reaction is calculated from information on the number of grams.

Write T *for true or* F *for false. If a statement is false, replace the underlined word or phrase with one that will make the statement true, and write your correction on the blank provided.*

______________________ 5. The coefficients in a balanced chemical equation show the correct ratio of masses.

______________________ 6. In solving a mass-mass problem, it is not necessary to work with a balanced chemical equation.

______________________ 7. In a reaction with a low percentage yield, the amount of product produced, compared with the amount that theoretically could have been produced, is small.

______________________ 8. The heat of reaction is represented by the letter *q*.

______________________ 9. The heat of reaction for a system giving off energy has a positive value.

______________________ 10. To convert grams to moles, the number of grams is multiplied by a factor that has the units mol/g.

______________________ 11. A chemical equation that shows an energy term on the left represents a reaction that absorbs energy.

Chapter 10

STUDY GUIDE

10.1 PERIODIC TRENDS

Write a definition for each of the following terms.

1. atomic radius

2. noble gas configuration

Answer each question in complete sentences.

3. Why does an atom of sodium have a larger atomic radius than an atom of chlorine has, even though sodium has fewer electrons?

4. When dissolved in water, NaCl will carry an electric current, but when the NaCl is in solid form, it will not do so. Explain this fact.

5. If iron has an electron configuration of $1s^22s^22p^63s^23p^64s^23d^6$, how does it form ions with oxidation numbers of 2+ and 3+?

6. Predict the most likely oxidation number for atoms that have the following electron configurations.
   a. $1s^22s^22p^63s^23p^64s^23d^{10}$
   b. $1s^22s^22p^63s^23p^5$
   c. $1s^22s^22p^63s^23p^1$
   d. $1s^22s^22p^63s^23p^64s^23d^{10}4p^65s^14d^{10}$

Name ______________________ Date ______________ Class ______________

Chapter 10

STUDY GUIDE

10.2 REACTION TENDENCIES

Write a definition for each of the following terms.

1. ionization energy ______________________________

2. first ionization energy ______________________________

3. shielding effect ______________________________

4. electron affinity ______________________________

Answer the following question in a complete sentence.

5. State the four factors that affect ionization energy, and briefly describe the effect of each.

____

Complete each of the following statements.

6. The first ionization energy tends to ______________ as atomic number increases in any horizontal row, or period.

7. A column, or group, will show a decrease in first ionization energy as atomic number ______________.

8. Ionization energy is typically measured in units called ______________.

9. Metals have a ______________ first ionization energy, and nonmetals have a ______________ first ionization energy.

10. Nonmetals have ______________ electron affinities. Metals have ______________ electron affinities. Moving down a column, the values of the elements' electron affinities tend to ______________.

Chapter 11

STUDY GUIDE

11.1 HYDROGEN AND MAIN GROUP METALS

Write T *for true or* F *for false. If a statement is false, replace the underlined word or phrase with one that will make the statement true, and write your correction on the blank provided.*

______________ 1. The shielding effect causes large atoms to lose their electrons more readily than they might otherwise.

______________ 2. A hydride ion is a bare proton.

______________ 3. A catalyst is a substance that speeds up a reaction.

______________ 4. By gaining an electron, hydrogen attains the stable electron configuration of helium.

______________ 5. As atoms of the alkali metals increase in size, they lose their outermost electron more easily.

______________ 6. Baking soda is another name for sodium carbonate.

______________ 7. The thallium ion and the ammonium ion each have a charge of 1+.

______________ 8. The alkali metal that is insoluble in most other alkali metals is cesium.

______________ 9. The alkali metals make up Group 2 of the periodic table.

______________ 10. Lime is a calcium compound.

______________ 11. Magnesium is the most plentiful metal in Earth's crust.

______________ 12. Alkaline earth metals tend to form 2+ ions.

Match each element with the correct description.

a. sodium
b. francium
c. hydrogen
d. potassium
e. magnesium
f. calcium
g. aluminum
h. beryllium
i. lithium
j. radium

____ 13. the most active metal
____ 14. is used in making nonsparking tools
____ 15. forms a silicate used as a catalyst and in soapmaking
____ 16. found in large amounts in muscle and nerve tissue
____ 17. can form both positive and negative ions
____ 18. its ions are a major constituent of bone and affect release and absorption of hormones
____ 19. the largest alkaline earth metal atom
____ 20. has three outermost electrons
____ 21. is part of a chlorophyll molecule
____ 22. alkali metal that reacts most vigorously with nitrogen and that burns in air to form an oxide, rather than a peroxide

Name ______________________ Date ______________ Class ______________

Chapter 11

STUDY GUIDE

11.2 NONMETALS

Complete each sentence.

1. When most of the elements in Group 14 (IVA) react, they tend to ______________ electrons.
2. Different forms of the same element are called ______________.
3. Dry ice is the name for solid ______________.
4. The Haber process produces ______________.
5. The transfer of genetic information from generation to generation involves a nitrogen-phosphorus organic compound called ______________.
6. The oxide ion has a charge of ______________.
7. The form of oxygen that has the formula O_3 is called ______________.
8. When SO_3 is dissolved in water, the compound called ______________ is formed.
9. Ions made up of long chains of sulfur atoms attached to the S^{2-} ion are called ______________ ions.
10. Chlorine and bromine are members of the ______________ family.
11. The least reactive elements make up a family called the ______________.
12. A mixture of metals is called a(n) ______________.
13. A mixture of copper and tin is called ______________.
14. When the members of Group 17 (VIIA) react, they usually ______________ electrons.
15. Going down Group 17 (VIIA) of the periodic table, there is a(n) ______________ in the activity of the elements.

Match each element with its description. Some choices may serve as answers more than once.

a. argon
b. oxygen
c. xenon
d. lead
e. fluorine
f. carbon
g. selenium
h. silicon
i. phosphorus
j. nitrogen
k. helium

____ 16. the most reactive nonmetallic element

____ 17. the most plentiful element in Earth's crust

____ 18. its allotropes are graphite and diamond

____ 19. the second most plentiful element in Earth's crust

____ 20. a distinctly metallic element in Group 14 (IVA)

____ 21. used in light bulbs to protect the filament

____ 22. a constituent of all organic compounds

____ 23. makes up most of Earth's atmosphere

____ 24. the first noble gas to produce a compound

____ 25. a semiconductor used in transistors and computer chips

____ 26. produces solder when mixed with tin

____ 27. in its white form, ignites spontaneously in air

____ 28. was first discovered somewhere other than on Earth

____ 29. a metalloid member of Group 16 (VIA)

____ 30. the only element to exhibit catenation to a great extent

Write T *for true or* F *for false. If a statement is false, replace the underlined word or phrase with one that will make the statement true, and write your correction on the blank provided.*

______ 31. Carbonic acid and cyanides are examples of organic compounds.

______ 32. Tin is an example of a metal that has been known since prehistoric times.

______ 33. Some bacteria convert atmospheric oxygen to compounds that can be used readily by plants to make amino acids.

______ 34. Elemental phosphorus occurs as P_2 molecules.

______ 35. A Group 15 (VA) element that occurs in all oxidation states from 3– to 5+ is nitrogen.

______ 36. Argon is an example of an element that has never been made to form compounds.

______ 37. Elements in the carbon family have complete outer energy levels.

Name ______________________ Date ______________ Class ______________

Chapter 11

STUDY GUIDE

11.3 TRANSITION METALS

Complete each sentence.

1. Steel is an alloy of ______________.
2. The highest-energy electrons of transition metals are in the ______________ sublevel.
3. There is a total of ______________ columns, or groups, of transition metals in the periodic table.
4. The CrO_4^{2-} ion is called the ______________ ion.
5. In most reactions, chromium loses ______________ of its electrons.
6. Iron is galvanized by being dipped into molten ______________.
7. Rubies and emeralds get their color from ______________ impurities.
8. ______________ is an alloy of copper and zinc.
9. The highest-energy electrons of the inner transition elements are in the ______________ sublevel.
10. The inner transition elements of Period 6 are called ______________.
11. The inner transition elements of Period 7 are called ______________.
12. The most stable ion for lanthanoids has a charge of ______________.
13. The lanthanoid element ______________ forms alloys with unusual conductivity and magnetic properties.
14. ______________ is a highly toxic actinoid.
15. A transition metal that is found in many proteins in biological systems is ______________.

Match each element with its use described below.

a. silver
b. molybdenum
c. cobalt
d. zinc
c. mangancsc
f. neodymium
g. chromium
h. curium
i. titanium
j. osmium

____ 16. forms a dioxide used as a white paint pigment

____ 17. is used in spark plugs

____ 18. is needed in the diet for proper functioning of the pancreas

____ 19. is used to harden pen points

____ 20. forms a dioxide used in batteries

____ 21. is used in coinage

____ 22. is used to plate steel to protect it from corrosion

____ 23. may be used in the future as the energy source in satellite nuclear generators

____ 24. has a radioactive isotope used in cancer treatment

____ 25. forms an oxide used in glass filters and lasers

Chapter 12

STUDY GUIDE

12.1 BOND FORMATION

1. Define electronegativity.

2. Electron affinity and electronegativity are both measures of an atom's attraction for electrons. What is the difference between them?

3. How does electronegativity vary as the atomic number of an element increases within the same period of the periodic table?

4. How is the strength of a bond between two elements in a molecule related to their electronegativities?

5. What is the difference between an ionic and a covalent bond?

6. How is the character of a bond (ionic or covalent) between two elements related to their electronegativities?

7. Referring to Table 12.1, Electronegativities, in your text, arrange the following compounds in order of increasing ionic character of their bonds: LiF, LiBr, KCl, KI.

8. Referring to Tables 12.1 and 12.3 in your text, classify each of the following bonds as either ionic (I) or covalent (C):

____ a. Al–O
____ b. Al–S
____ c. Bi–Cl
____ d. Bi–O
____ e. C–Cl
____ f. N–O
____ g. Na–S
____ h. P–O
____ i. S–O
____ j. Ti–Br

Name ______________________ Date ______________ Class ______________

9. What force holds the two ions together in an ionic bond?

____

10. What is the meaning of the oxidation number of an element that forms an ionic bond?

____

____

11. What is a molecule?

____

12. What causes the bond lengths and bond angles of a molecule to vary?

____

13. a. Why does a molecular compound absorb specific frequencies of infrared radiation?

____

____

b. How can the absorption of this radiation be used to identify a compound?

____

____

14. Why are the bonding electrons in metallic bonding said to be delocalized?

____

____

15. What is one factor that determines the hardness of a metallic element?

____

____

16. Indicate whether each property listed below is characteristic of ionic (I), covalent (C), or metallic (M) bonding. More than one letter may be used for each answer.

____ a. Shape of solid can be changed by pounding.

____ b. Not electrically conducting in solid phase

____ c. Electrically conducting in all phases

____ d. High melting points

17. a. How are a molecule and a polyatomic ion alike?

____

b. How are they different?

____

Chapter 12

STUDY GUIDE

12.2 *PARTICLE SIZES*

1. a. Which is larger, a metallic atom or its positive ion?

   ______

   b. Which is larger, a nonmetallic atom or its negative ion?

   ______

2. In an ionic crystal, how can the internuclear distance between two ions be calculated?

   ______

   ______

3. In a molecule, how can the approximate bond length between two atoms be calculated?

   ______

   ______

4. Use Table 12.5 in your textbook to calculate the expected bond length (in picometers) of the following covalent bonds:

   ______ a. P–O

   ______ b. N–O

   ______ c. N–N

5. Arrange the atomic radius, covalent radius, and ionic radius of a nonmetallic atom in the expected order of increasing size.

   ______

6. Under what circumstances might the actual order of these sizes change?

   ______

   ______

   ______

   ______

Chapter 13

STUDY GUIDE

13.1 BONDS IN SPACE

Complete the sentence or answer the question.

1. Pairs of electrons that bond two atoms in a molecule are called ______________________.

2. Why do electron pairs in the outer levels of atoms in a molecule spread apart as far as possible?

____

3. Because each of the bond angles of methane equals 109.5°, its molecular shape is a perfect

______________________.

4. Complete the following diagrams to show the electron content of the outer orbitals of (a) an unbonded carbon atom, and (b) a hybridized carbon atom.

Unbonded carbon atom
Outer orbitals

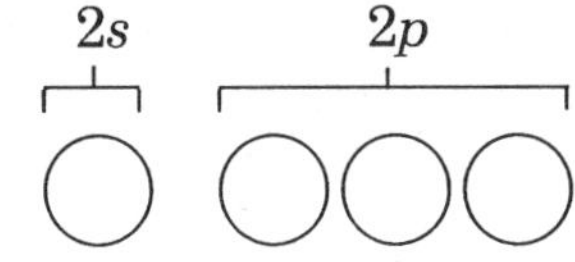

Hybridized carbon atom
Outer orbitals

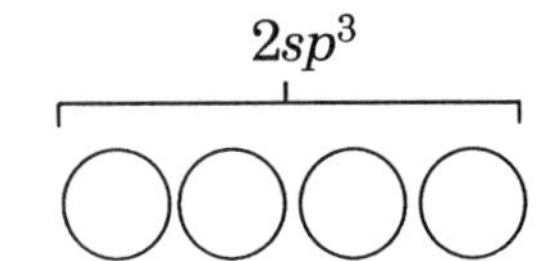

5. How do hybridized orbitals in an atom compare with one another?

____

____

6. When an sp^3 orbital of one carbon atom overlaps that of another carbon atom, how many electrons do the two atoms share?

____

7. How many hybrid orbitals of the following types can an atom have?

   a. sp^3 ____

   b. sp^2 ____

   c. sp ____

8. In a molecule of ethylene, C_2H_2, how many sigma bonds and how many pi bonds are present?

____

9. In a trigonal planar structure such as the CH_2O molecule, with one double bond and two single bonds, how does the double bond affect the angle between the two single bonds?

____

____

Name ____________________ Date ____________________ Class ____________________

Write T *for true or* F *for false. If a statement is false, replace the underlined word or phrase with one that will make the statement true, and write your correction on the blank provided.*

______________ 10. A <u>shared pair</u> of electrons is formed when an orbital of one atom overlaps an orbital of another atom.

______________ 11. <u>Sigma</u> bonds are always formed by the sideways or parallel overlap of unhybridized *p* orbitals.

______________ 12. In single, double, and triple bonds between carbon atoms, one of the bonds is always a <u>pi</u> bond.

______________ 13. When two carbon atoms are joined by more than one bond, the additional bonds are always <u>pi</u> bonds.

______________ 14. Pi bonds break more easily than do sigma bonds, because the electrons forming pi bonds are <u>closer to</u> the nuclei.

Chapter 13

STUDY GUIDE

13.2 MOLECULAR ARRANGEMENTS

Complete the sentence or answer the question.

1. In saturated organic compounds, all the bonds between carbon atoms are ______.

2. The existence of two or more substances with the same molecular formula but different structural formulas is known as ______.

3. What is the difference between structural isomerism and positional isomerism?

______

______

4. Compounds with the same atoms bonded in the same order but with a different arrangement of atoms around a double bond are examples of ______.

5. a. What type of hybridization would be expected in an atom with three outer electrons?

   ______

   b. What would be the shape and bond angles of the molecule it forms?

   ______

Match each suffix with its correct description of the bonding of the carbon atoms in a hydrocarbon.

a. saturated compounds (single bonds only)
b. unsaturated compounds with double bonds
c. unsaturated compounds with triple bonds

____ 6. -yne

____ 7. -ane

____ 8. -ene

Match each of the following compounds with the correct description of its molecule.

a. boron trichloride (BCl_3)
b. acetylene (ethyne) (C_2H_2)
c. ozone (O_3)
d. oxygen difluoride (OF_2)
e. carbon tetrachloride (CCl_4)
f. trimethylarsine $(CH_3)_3As$

____ 9. linear, with 180° bond angles

____ 10. bent, with an exceptionally small bond angle resulting from two unshared pairs of electrons

____ 11. bent, with a small bond angle resulting from one unshared pair of electrons

____ 12. tetrahedral, with the expected bond angles of 109.5°

____ 13. trigonal pyramidal, with an unusually small bond angle (96°)

____ 14. trigonal planar, with the expected bond angles of 120°

For each of the following pairs of isomers, describe the difference in structure and identify the type of isomerism illustrated.

15.

```
    H H H             H  OH H
    | | |             |  |  |
  H-C-C-C-OH        H-C - C - C-H
    | | |             |  |  |
    H H H             H  H  H
```

1-propanol   2-propanol

Structural difference: _______________

Type of isomerism: _______________

16.

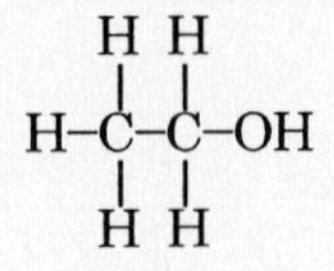

ethanol

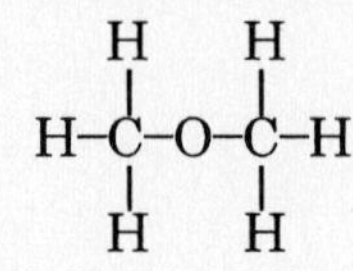

methoxymethane

Structural difference: _______________

Type of isomerism: _______________

Complete the sentence or answer the question.

17. a. In the compound named ethene, what does the ending *-ene* indicate about its structure?

____

____

b. What does the stem *eth-* indicate?

____

____

18. a. Draw the structural formula for $CH\equiv C–CH_2–CH_3$.

b. Give the name of this compound, and explain the meaning of its stem and ending.

____

____

____

19. Under each of the following simplified structural diagrams, write the name of the compound represented.

 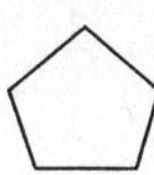  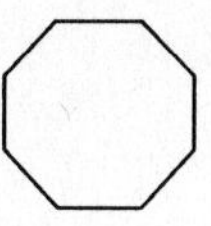

20. Draw the structural formula for the positional isomer of the compound in question 18.

Chapter 14

STUDY GUIDE

14.1 MOLECULAR ATTRACTION

Complete the sentence or answer the question.

1. Define the following terms.
   a. polar covalent bond ________________

   ____

   b. dipole ________________

   ____

   c. dipole moment ________________

   ____

2. Are the polar bonds in a polar molecule arranged symmetrically or asymmetrically?

   ____

3. What does the dipole moment of a molecule tell us about its intermolecular forces?

   ____

4. Indicate whether the following molecules are polar or nonpolar.
   a. BeF_2 ________________
   b. H_2O ________________
   c. $CHCl_3$ ________________
   d. CCl_4 ________________

Write the letter of the term that matches the correct description.

a. van der Waals forces
b. intramolecular forces
c. intermolecular forces
d. dipole-dipole forces
e. dipole-induced dipole forces
f. temporary dipole
g. dispersion forces
h. induced dipole

____ 5. a nonpolar molecule in which the charge distribution is briefly asymmetrical

____ 6. a nonpolar molecule that has its electron cloud distorted by an approaching dipole and is thus transformed into a dipole

____ 7. weak forces involving the attraction of the electrons of one atom for the protons of another

____ 8. attractive forces between two molecules of the same or different substances that are both permanent dipoles

____ 9. forces generated by temporary dipoles

___ 10. forces between molecules

___ 11. attractive forces between dipoles and nonpolar molecules

___ 12. forces within a molecule

Write T *for true and* F *for false. If a statement is false, replace the underlined word or phrase with one that will make the statement true, and write your correction on the blank provided.*

______________ 13. Dispersion forces are the only attractive forces that attract between nonpolar molecules.

______________ 14. Molecules cannot exhibit both dipole and dispersion interactions.

______________ 15. Substances composed of nonpolar molecules are generally gases at room temperature or high-boiling liquids.

______________ 16. Substances composed of polar molecules generally have higher boiling points than do nonpolar compounds.

______________ 17. A nonpolar molecule can have polar bonds.

______________ 18. Water is a nonpolar molecule that has polar bonds.

Name ______________________ Date ______________ Class ______________

Chapter 14

STUDY GUIDE

14.2 COORDINATION CHEMISTRY

Complete the sentence or answer the question.

1. What is a complex ion?

____

2. Name two uses for complex ions.

____

3. A ligand is either a ______________ or a ______________ that is attached to a central positive ion in a complex ion.

4. The coordination number of a complex ion is the number of ______________ of ligands surrounding the central positive ion.

5. Why is the oxalate ion in a complex ion called a didentate ligand?

____

6. What type of complex is formed by three didentate ligands? ______________

7. The molecules of a coordination compound are complexes with a net charge of

______________ .

Write T *for true and* F *for false. If a statement is false, replace the underlined word or phrase with one that will make the statement true, and write your correction on the blank provided.*

______________ 8. In a complex ion, <u>nonpolar</u> molecules or negative ions are attached to a central positive ion.

______________ 9. Ligands can be either molecules or <u>positive</u> ions.

______________ 10. The most common ligand is <u>water</u>.

______________ 11. Complex ions with coordination number <u>two</u> are always linear.

______________ 12. In a coordinate covalent bond the electrons in the shared pair come from <u>different atoms</u>.

______________ 13. Although the bonds of most complex ions have characteristics of both covalent and ionic bonding types, the <u>ionic</u> character dominates.

14. Complete the following table.

Coordination number	Shape of complex
2	
4	
6	

Name ______________________ Date ______________ Class ______________

Chapter 14

STUDY GUIDE

14.3 CHROMATOGRAPHY

Complete the sentence or answer the question.

1. Define the following terms.

   a. fractionation ______________________

   b. chromatography ______________________

   ____

2. Which phase in chromatography consists of the mixture to be separated being dissolved in a fluid (liquid or gas)? ______________

3. Label the following parts of the gas chromatography system in the figure.

   a. inert carrier gas
   b. stationary phase
   c. mobile phase
   d. volatile mixture to be separated

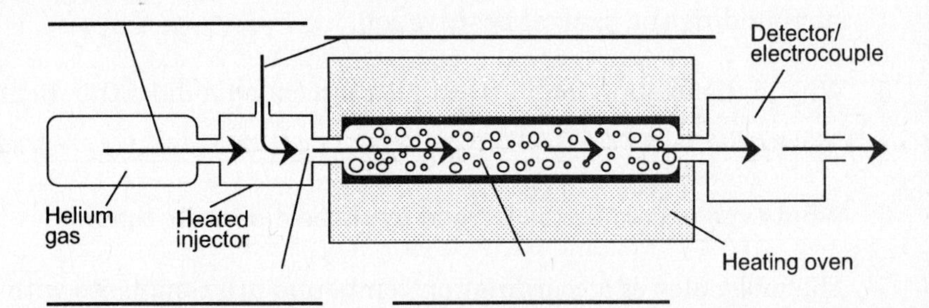

Write T *for true or* F *for false. If a statement is false, replace the underlined word or phrase with one that will make the statement true, and write your correction on the blank provided.*

______________ 4. Chromatography is a method of separating a mixture of substances into components based on differences in polarity of the components.

______________ 5. In chromatography, the fastest migrating substance will be the one with the least attraction for the stationary phase.

______________ 6. Paper chromatography is used when it is necessary to speed up the separation process.

______________ 7. An ion exchange resin is used as the stationary phase of thin layer chromatography.

Write the letter of the type of chromatographic technique that matches the description.

a. column chromatography
b. high performance liquid chromatography
c. ion chromatography
d. paper chromatography
e. thin layer chromatography
f. gas chromatography

____ 8. combines some of the techniques of both column and paper chromatography and is used frequently in separating biological materials

____ 9. is a chromatographic technique for the analysis of volatile liquids and mixtures of gases

____ 10. is designed to overcome the speed limitations of conventional column chromatography

____ 11. is a simple and fast chromatographic technique, but one in which quantitative determinations are difficult because of its extremely small scale

____ 12. is used for vitamins, proteins, and hormone separations not easily made by other methods

Chapter 15

STUDY GUIDE

15.1 PRESSURE

Write T for true or F for false. Is a statement is false, replace the underlined word or phrase with one that will make the statement true, and write your correction in the blank.

______________ 1. All matter is composed of <u>small particles</u>.

______________ 2. Particles that make up matter are always <u>stationary</u>.

______________ 3. Collisions between particles of matter are perfectly elastic because there is <u>no change</u> in the total kinetic energy of the particles before and after the collision.

______________ 4. Under ordinary conditions, each molecule of a gas undergoes a few <u>million</u> collisions each second.

______________ 5. A barometer is <u>an open-arm</u> manometer used to measure atmospheric pressure.

______________ 6. Standard atmospheric pressure is <u>one pascal</u>.

Complete the statement or answer the question.

7. What is one pascal?

____

8. In using any manometer, what measurement must you make?

____

9. What physical property of the liquid in a manometer must you know?

____

10. Why is the measurement of gas pressure with a closed-arm manometer independent of the atmospheric pressure?

____

11. Standard atmospheric pressure will support a column of mercury ______________ mm high.

12. One kilopascal equals ______________ mm Hg.

13. A closed-arm mercury manometer is shown being used to measure the pressure of a gas in a closed container. What is the pressure of the gas in kilopascals?

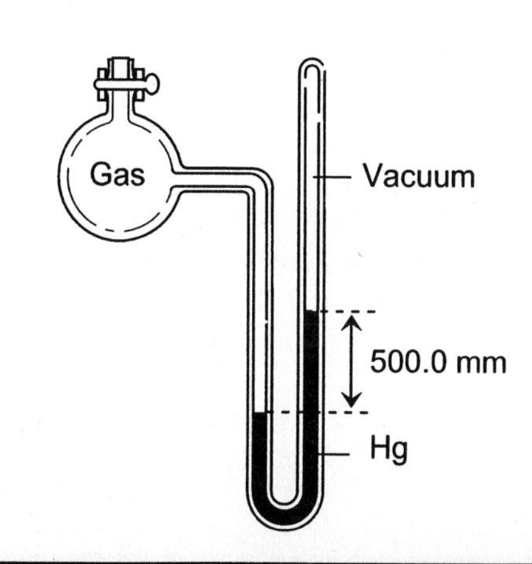

Name ______________________ Date ______________ Class ______________

Chapter 15

STUDY GUIDE

15.2 MOTION AND PHYSICAL STATES

Answer the following questions.

1. Complete the following table comparing the three common states of matter.

State of matter	Shape	Volume
solid		definite
	assumes the shape of its container	
gas		

2. What two factors determine the average speed of the particles in a gas?

____

3. Define *absolute zero.*

____

4. In which direction does energy flow between two objects at different temperatures?

____

5. What is the SI unit of temperature?

____

6. What is absolute zero on the Celsius temperature scale?

____

7. What equation can be used to convert Celsius temperatures (°C) to Kelvin temperatures (K)?

____

8. Convert the following temperatures to the Kelvin scale.
   a. 15°C ______________
   b. –100°C ______________
   c. 0°C ______________

9. Convert the following temperatures to the Celsius scale.
   a. 0 K ______________
   b. 100 K ______________
   c. 722 K ______________

10. What is the difference between heat and temperature?

____

____

Name ______ Date ______ Class ______

Chapter 16

STUDY GUIDE

16.1 *CRYSTAL STRUCTURE*

1. Complete the table.

Crystal system	Lengths of unit cell axes	Angle between unit cell axes
cubic	equal	
tetragonal		all = 90°
	all unequal	all = 90°
	all unequal	2 = 90°, 1 ≠ 90°
triclinic		
rhombohedral		
hexagonal		1 = 90°, 3 = 60°

2. Identify each of the crystal shapes in Figure 1.

Figure 1

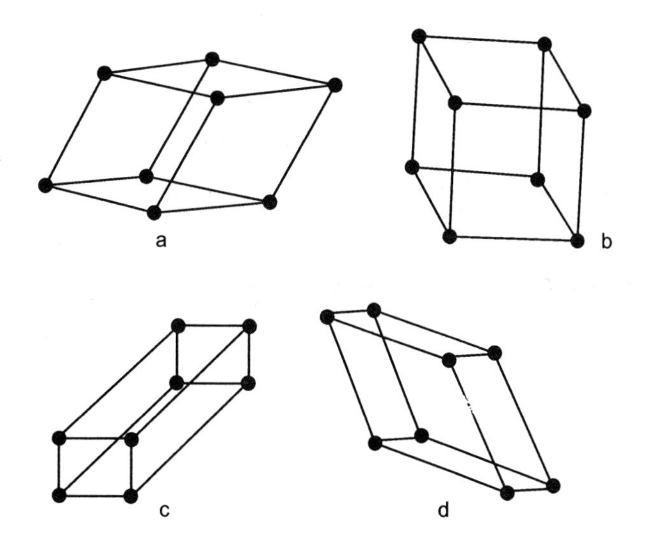

a. ______

b. ______

c. ______

d. ______

Name ____________ Date ____________ Class ____________

3. Identify each of the unit cells in Figure 2.

Figure 2

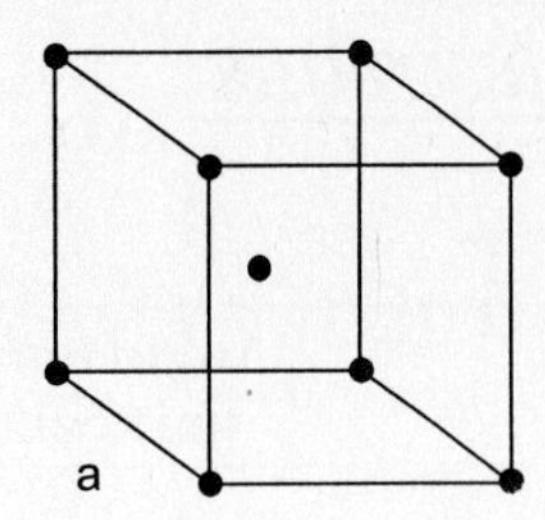

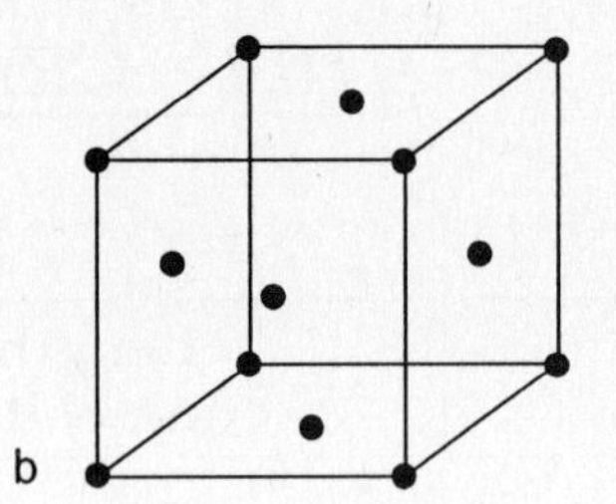

a. ____________

b. ____________

Figure 3

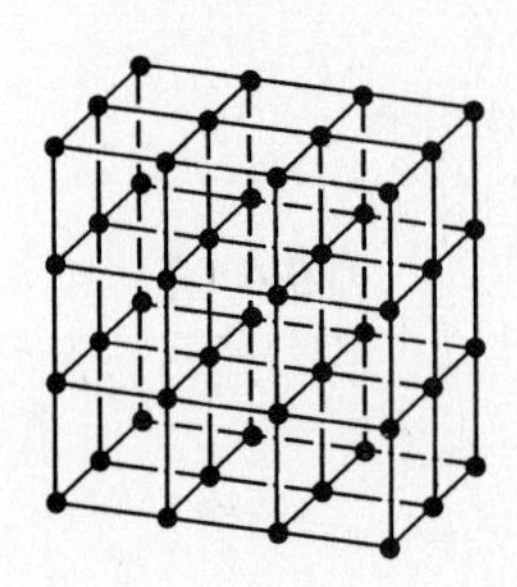

4. a. What is the general name for a structure like the one shown in Figure 3?

____

b. If the unit cell is simple cubic, how many unit cells does the structure in Figure 3 contain?

____

Match each type of bonding with the correct type of crystal.

a. covalent
b. delocalized electrons
c. electrostatic forces
d. van der Waals forces

____ 5. ionic

____ 6. metallic

____ 7. macromolecular/network

____ 8. molecular

9. Label each range of melting points on the temperature scale in Figure 4 with the name of the appropriate crystal type.

Figure 4

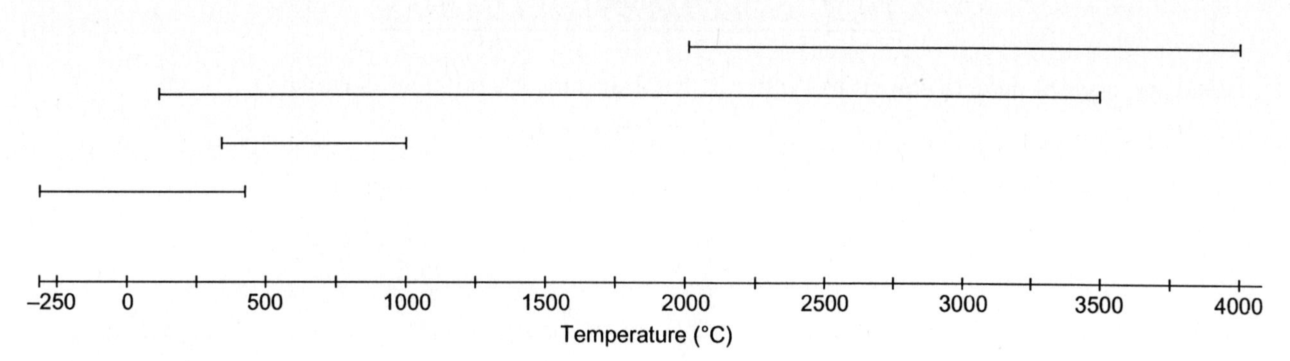

Write T *for true or* F *for false. If a statement is false, replace the underlined word or phrase with one that will make the statement true, and write your correction on the blank provided.*

_______________ 10. In a crystal of sodium chloride, each Na^+ ion is surrounded by <u>four</u> Cl^- ions.

_______________ 11. The unit cell of sodium chloride is <u>body-centered</u> cubic.

_______________ 12. Crystals of different solids with the same structure and shape are said to be <u>polymorphous</u>.

Name ______________________ Date ______________ Class ______________

Chapter 16

STUDY GUIDE

16.2 SPECIAL STRUCTURES

1. Label the crystal defects shown in Figure 1. Explain how each defect is caused.

Figure 1

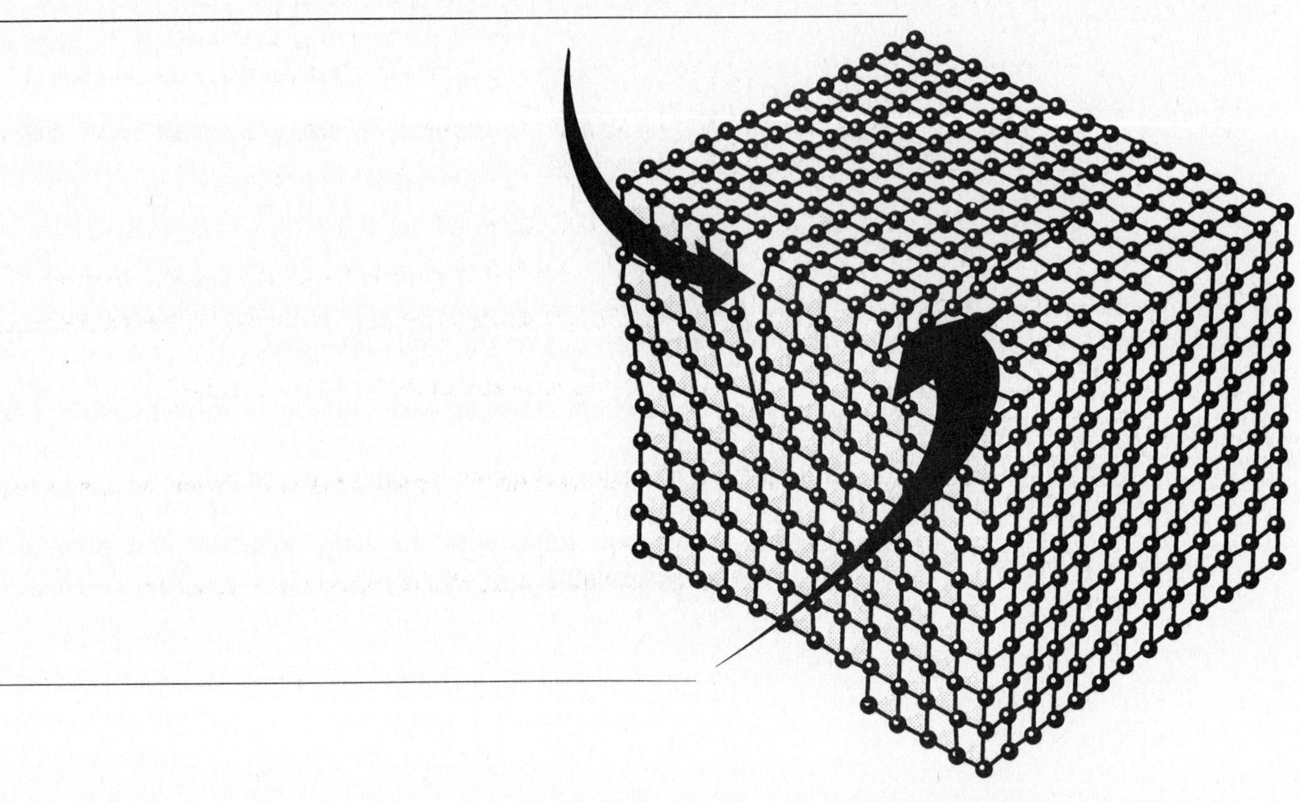

Write T *for true or* F *for false. If a statement is false, replace the underlined word or phrase with one that will make the statement true, and write your correction on the blank provided.*

______________ 2. Imperfect crystals can be caused by defects to the unit cell, such as misplaced atoms or ions.

______________ 3. Hydrated ions are chemically bonded to water molecules.

______________ 4. $CuSO_4 \cdot 5H_2O$ and $CaSO_4 \cdot 2H_4O$ are anhydrous compounds.

______________ 5. It is possible to remove water molecules from hydrated compounds by raising the temperature or lowering the pressure.

______________ 6. Deliquescent materials are the least hygroscopic substances.

______________ 7. Materials that appear to be solid but that lack crystalline structures are called amorphous materials.

______________ 8. The symbol for amorphous materials is (amph).

______________ 9. Glass and molasses have lower viscosities than water and alcohol because they tend to flow less easily.

______________ 10. Metastable forms of an amorphous substance occur in long-lasting form.

Use the letters of the two-dimensional diagrams of (a) crystalline quartz and (b) quartz glass, shown in Figure 2, to answer the following questions.

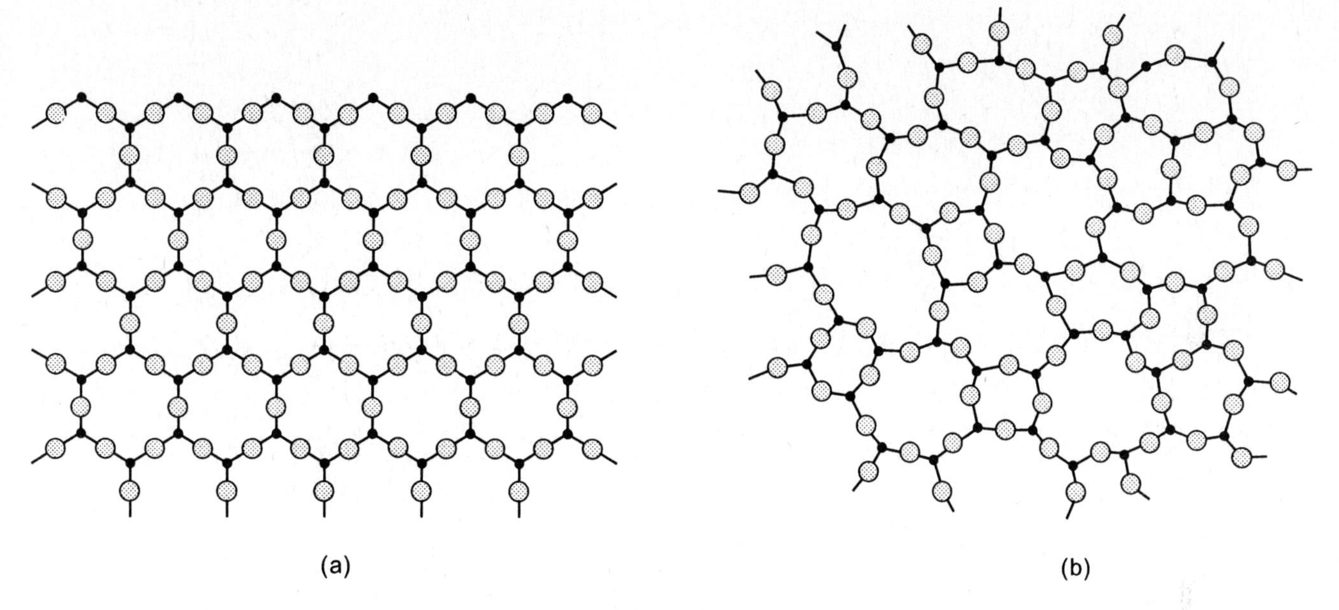

Figure 2

____ 11. Which substance shows the structure of an amorphous material?

____ 12. Which substance illustrates long-range order?

____ 13. Which substance has a specific melting point?

____ 14. Which substance can be classified as a supercooled liquid?

____ 15. Which substance is less stable in the form shown?

Name ______________________ Date ______________________ Class ______________________

Chapter 17

STUDY GUIDE

17.1 CHANGES OF STATE

Complete the sentence or answer the question.

1. What is formed when molecules of a liquid or a solid substance escape from the surface of the substance? ______________________

2. According to kinetic theory, if the temperature of a liquid is lowered, the average velocity of its particles should ______________________.

3. When the ordered arrangement of the atoms of a solid substance breaks down, the solid ______________________.

4. When a liquid ______________________ the particles of the substance settle into an ordered arrangement and form a solid.

5. The equation $X(l) \rightarrow X(g)$ represents change from ______________________.

6. Write an equation that shows water, H_2O, as it changes reversibly between the liquid and the solid states. ______________________

7. According to Le Chatelier's principle, what happens if stress is applied to a system at equilibrium?

____

8. Write the name of the device shown in the figure.

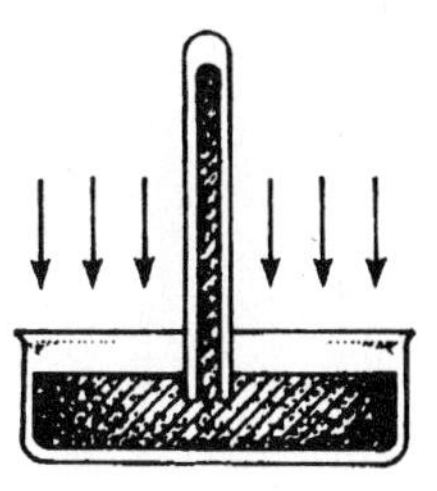

____

9. What are the two devices shown in Figure 17.1 used to measure? ______________________

10. What is true about the vapor pressures of substances that have strong intermolecular forces?

____

11. What is meant by the melting point for a substance?

____

12. When dry ice is placed in an open container at room temperature, it changes directly from a solid to a vapor. This process is an example of ______________________.

13. The normal boiling point of a substance is the temperature at which its vapor pressure is equal to

____

Name ______________________ Date ______________ Class ______________

14. A liquid that boils at a low temperature and evaporates rapidly at room temperature is described as being ______________.

15. The temperature above which no amount of pressure will liquefy a gas is the ______________ of the gas.

16. The condensation of substances that are normally gases is called ______________.

17. Placing a checkmark in the correct column, identify whether each characteristic listed in the table applies to a substance with strong intermolecular forces or to a substance with weak intermolecular forces.

Intermolecular forces		
Characteristic	**Strong**	**Weak**
Volatile		
High boiling point		
High evaporation rate		
Low vapor pressure at room temperature		
Low critical temperature		

18. The specific heat of ice is 2.06 J /g·C°. Calculate how much energy is needed to heat a 52.0-g sample of ice from –20.0°C to 0.0°C.

19. The enthalpy of fusion of ice is 334 J/g. Calculate how much energy is needed to melt a 52.0-g sample of ice.

20. The C_p of liquid water is 4.18 J/g·C°. Calculate how much energy is needed to heat a 52.0-g sample of liquid water from 0.0°C to 100.0°C.

21. The enthalpy of vaporization of water is 2260 J/g. Calculate how much energy is needed to boil a 52.0-g sample of liquid water that is already at the boiling point.

22. The C_p of steam is 2.02 J/g·C°. Calculate how much energy is needed to heat 52.0-g of steam from 100.0°C to 120.0°C.

23. Use your calculations from questions 18 through 22 to calculate how much energy is necessary to convert a 52.0-g sample of ice at –20.0°C to steam at 120.0°C.

Name ______________________ Date ______________________ Class ______________________

Chapter 17

STUDY GUIDE

17.2 SPECIAL PROPERTIES

Write T *for true and* F *for false. If a statement is false, replace the underlined word or phrase with one that will make the statement true, and write your correction on the blank provided.*

______________________ 1. Water contracts when it freezes.

______________________ 2. Compared with most molecules, water molecules have low masses.

______________________ 3. Water is a gas at room temperature and standard atmospheric pressure.

______________________ 4. Carbon dioxide and nitrogen are gases at room temperature and standard atmospheric pressure.

______________________ 5. The structure of the molecules in a substance affects the interatomic and intermolecular forces that hold the substance together.

______________________ 6. Molecules that contain hydrogen that is covalently bonded to a highly electronegative atom often have boiling points and melting points that are higher than would be expected in the absence of such hydrogen.

______________________ 7. A molecule that is made up of a highly electronegative atom and a hydrogen atom is highly polar.

______________________ 8. If an actual H^+ ion existed, it would consist of a bare neutron.

______________________ 9. In compounds, hydrogen is always ionically bonded.

______________________ 10. In a substance that is made up of identical polar molecules that contain a hydrogen atom, the hydrogen atom of each molecule in the substance is attracted to the positive portion of the other molecules.

______________________ 11. A hydrogen bond results in a fairly strong hydrogen atom attachment between two molecules.

______________________ 12. The attractive force between hydrogen-bonded molecules is much greater than the attractive force between other dipoles with the same electronegativity difference.

______________________ 13. Hydrogen bonding is one type of dipole attraction.

______________________ 14. Liquid water at 1°C is less dense than liquid water at 3°C.

______________________ 15. Because ice is more dense than liquid water, ice floats at the surfaces of lakes and streams.

Match each term with the correct description.

a. hydrogen bonding
b. capillary rise
c. surface tension

___ 16. apparent elasticity at the surface of a liquid

___ 17. change in elevation of a liquid in a tube with a small diameter

___ 18. an attractive force that exists between molecules that contain hydrogen and a highly electronegative atom

ame ______________________ Date ______________ Class ______________

19. A water molecule is a dipole. Look at the diagram of the water molecule. Place + and – signs in the blanks to label the areas of the molecule that are partially positive and partially negative.

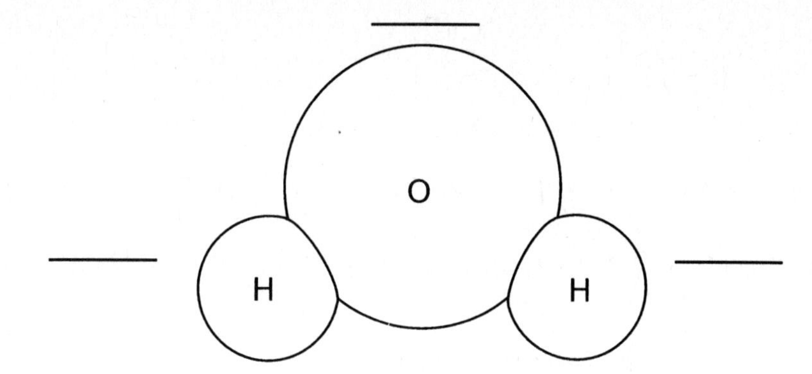

Chapter 18

STUDY GUIDE

18.1 VARIABLE CONDITIONS

Complete the sentence or answer the question.

1. According to the kinetic theory, a gas is made of very small particles that are in constant random ________.

2. What are the three factors on which the pressure exerted by a gas depends?

____

____

3. a. If the number of molecules in a constant volume increases, the pressure ________.
   b. If the number of molecules and the volume remain constant, but the kinetic energy of the molecules increases, the pressure ________.
   c. The kinetic energy depends on the ________.

4. Study the graph and answer the questions.
   a. What gas law is illustrated in the graph?

   ____

   b. Write a sentence describing the relationship represented by the graph.

   ____

   ____

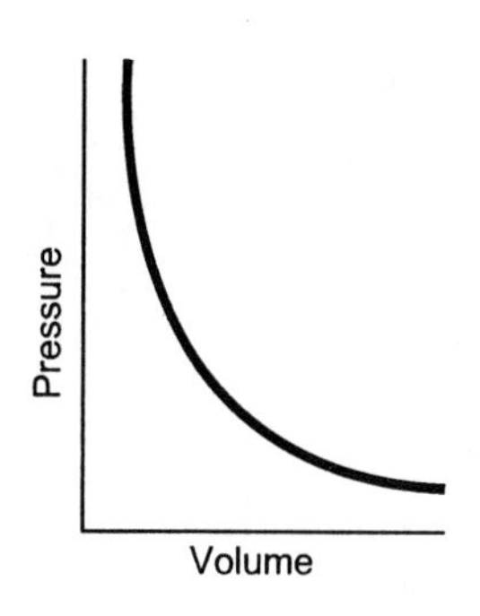

Write T *for true or* F *for false. If a statement is false, replace the underlined word or phrase with one that will make the statement true, and write your correction on the blank provided.*

________ 5. When a gas is one in a mixture of gases, the pressure exerted by the individual gas is called its <u>ideal</u> pressure.

________ 6. In calculations involving gases, the temperature given in Celsius must be converted to <u>Fahrenheit</u>.

________ 7. At a constant pressure, the volume of a quantity of gas varies <u>directly</u> with the Kelvin temperature.

________ 8. Gases collected by water displacement must be <u>soluble</u> in water.

________ 9. <u>Charles's</u> law is represented by $P_{total} = P_1 + P_2 + \ldots P_n$.

Use Tables 18.1 and 18.2 in the text to answer questions 10 and 11.

10. What is the partial pressure of oxygen in the air at standard conditions?

11. a. What is the vapor pressure of water at 32°C? ________
    b. At what temperature is the vapor pressure of water 2.2 kPa? ________

12. Complete the table by using Boyle's law. In the column labeled *Change in volume,* indicate whether the volume increases or decreases. In the next column, write the two possible ratios for pressure and circle the appropriate ratio needed for calculations. Calculate the final volume.

Initial volume	Change in pressure	Change in volume	Pressure ratios	Final volume
26 cm^3	55.8 kPa to 110.1 kPa			
131 dm^3	225 kPa to 650.5 kPa			
88 dm^3	36.8 kPa to 22.4 kPa			
925.0 cm^3	151.2 kPa to 119.7 kPa			
0.621 m^3	49.5 kPa to 76.2 kPa			

13. A 400-cm^3 volume of gas is collected at 26°C. What volume would this gas occupy at standard conditions? Assume a constant pressure.

Match each equation below with the correct descriptions.

a. $P_{gas} = P_{total} - P_{water}$ b. $PV = k$ c. $P_{total} = P_1 + P_2 + \ldots P_n$ d. $V = k'T$ e. K = °C + 273.15

____ 14. relationship between the Kelvin and Celsius temperature scales

____ 15. Boyle's law

____ 16. Charles's law

____ 17. Dalton's law

____ 18. used to find pressure of a gas collected over water

ame ______________________ Date ______________ Class ______________

Chapter 18

STUDY GUIDE

18.2 *ADDITIONAL CONSIDERATIONS OF GASES*

Answer the questions in the spaces provided.

1. Why is it usually necessary to correct laboratory volumes of gas for both temperature and pressure?

____

____

2. How are the corrections made?

____

____

3. Why can the corrections for temperature and pressure be made in either order in one equation?

____

____

4. Why would people across the room from a newly opened bottle of perfume soon be able to smell the perfume?

____

____

5. Based on the diagrams in Figure 1, which gas diffuses faster, ammonia (top row) or carbon dioxide (bottom row)? ____

Figure 1

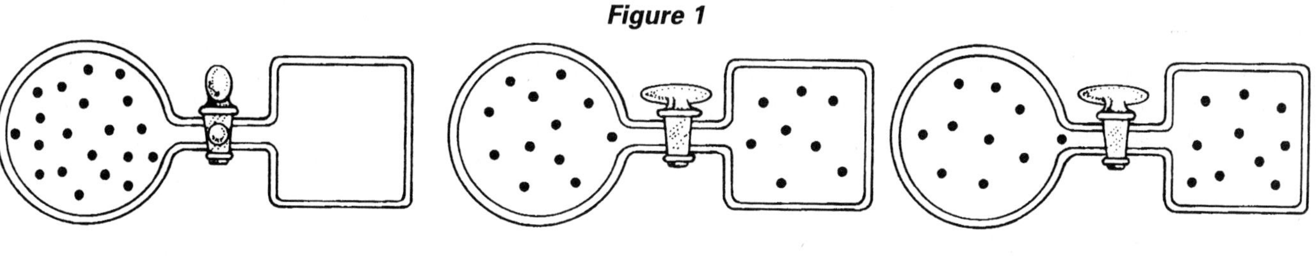

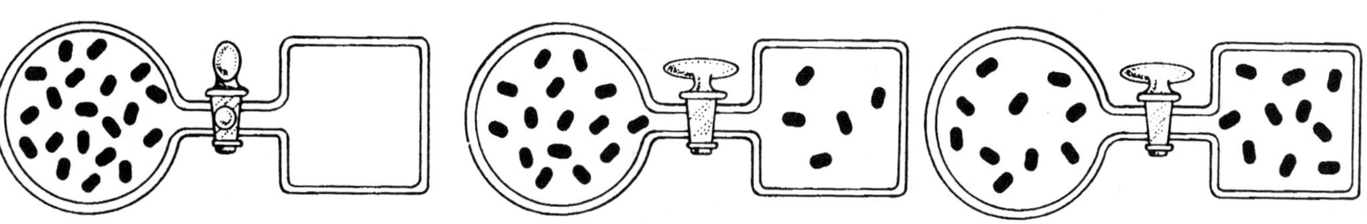

6. Calculate the relative rate of diffusion of the two gases.

Name ______________________ Date ______________________ Class ______________________

7. a. Which gas diffuses faster, helium or neon?

____

b. How much faster?

8. Using the illustrations for ammonia and carbon dioxide as guides, fill in the blank diagrams in Figure 2 comparing the rates of diffusion of helium and neon. Label your diagrams.

Figure 2

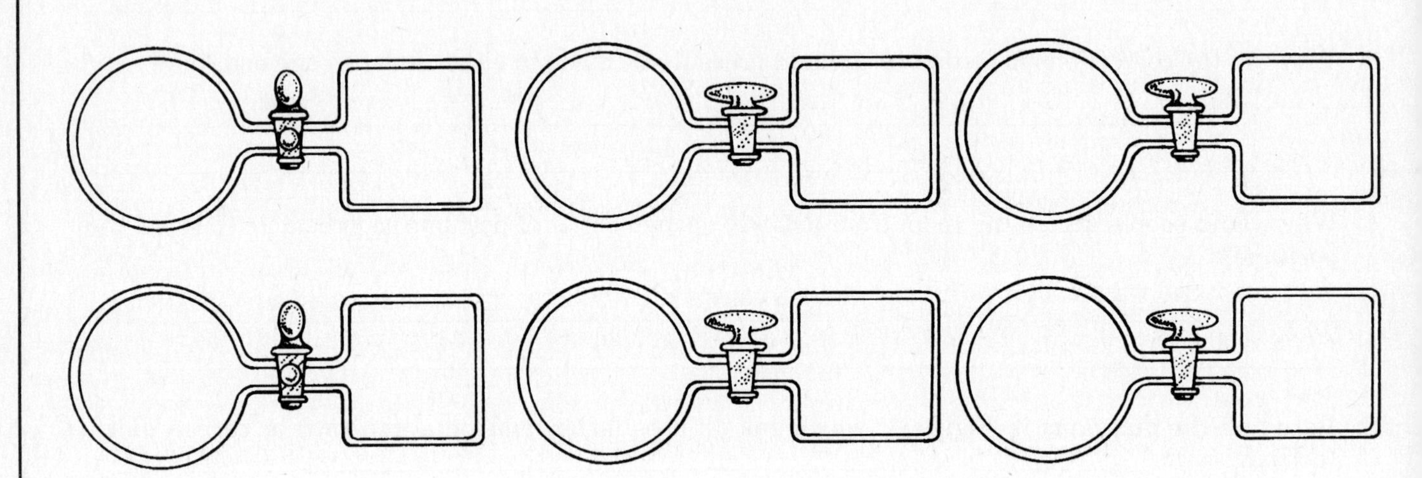

Chapter 19

STUDY GUIDE

19.1 AVOGADRO'S PRINCIPLE

Complete the sentence or answer the question.

1. What is Avogadro's principle?

______

______

2. Based on Avogadro's principle, if the volumes of two gases under similar conditions are equal, how are their number of moles related?

______

3. What is molar volume?

______

4. State what each of the five symbols in the ideal gas equation stands for.

______

______

5. What temperature scale is used to solve problems involving the ideal gas equation?

______

6. Using the ideal gas equation, complete the table by solving for the unknown variable. Show your work below.

P (kPa)	V (dm^3)	n	T (K)
55.6	85.0	6.50	
	210.0	3.20	293
113	8.5		311
95.1		5.50	277

7. In the modified form of the ideal gas equation $PV = mRT/M$, what do the symbols m and M stand for?

____

8. How can this modified equation also be used to solve for the density of a gas?

____

9. Complete the table by solving for M. Show your work below.

P (kPa)	V (dm^3)	m (g)	T (K)	M (g/mol)
98.6	6.27	0.832	300.0	
112.5	10.1	2.36	298	
88.7	23.5	0.55	292.5	
91.2	2.75	0.124	312	

Write T *for true or* F *for false. If a statement is false, replace the underlined word or phrase with one that will make the statement true, and write your correction on the blank provided.*

__________ 10. At a given temperature, the average kinetic energy of all gas molecules is the same.

__________ 11. One gram of oxygen will occupy 22.4 dm^3 at STP.

__________ 12. The Celsius temperature scale is used for problems involving the ideal gas equation.

__________ 13. Density is equal to mass divided by volume.

__________ 14. The ideal gas equation combines Boyle's law and Charles's law.

Chapter 19

STUDY GUIDE

19.2 GAS STOICHIOMETRY

Complete the sentence or answer the question.

1. What are the four steps you should keep in mind when solving mass-gas volume problems?

2. What are the four steps you should keep in mind when solving gas volume-mass problems?

Solve each of the following problems.

3. What volume of carbon dioxide, CO_2, at STP can be produced when 8.0 grams of oxygen, O_2, react with an excess of ethane, C_2H_6? Use the steps you listed in questions 1 and 2.

4. a. How many grams of carbon dioxide, CO_2, are formed if 96.0 g of oxygen, O_2, react with 12.2 dm^3 of ethylene, C_2H_4, to form carbon dioxide and water? Assume STP. Which reactant is the limiting one?

b. How many grams of CO_2 are produced when 20.0 g of O_2 react with 31 dm^3 of ethylene? Assume STP. Which reactant is the limiting one?

Write T for true or F for false. If a statement is false, replace the underlined word or phrase with one that will make the statement true, and write your correction on the blank provided.

______________________ 5. There is more than one way to find out which reactant is the <u>limiting reactant</u> in a chemical reaction.

______________________ 6. <u>Volume-volume</u> relationships can be used when all the reactants and products are gases.

______________________ 7. It is usually awkward to measure the <u>mass</u> of a gas.

______________________ 8. A balanced equation tells you the <u>volume</u> ratios of the reactants and products.

______________________ 9. In a chemical reaction, the amount of products is determined by the <u>excess</u> reactant.

Solve the following problems.

10. In a reaction between hydrogen and bromine to form hydrogen bromide gas, HBr, how many dm^3 of H_2 will be needed to completely use up 65.0 dm^3 of Br_2? (Assume constant pressure and temperature.)

11. How many moles and how many grams of HBr will be produced when the hydrogen and bromine react?

12. If only 22.4 dm^3 of H_2 reacted with 65 dm^3 of Br_2, how many grams of HBr would be produced?

ame ______________________ Date ______________ Class ______________

Chapter 20

STUDY GUIDE

20.1 SOLUTIONS

Write T for true or F for false. If a statement is false, replace the underlined word or phrase with one that will make the statement true, and write your correction on the blank provided.

______________ 1. Most solutions consist of a gas dissolved in a liquid.

______________ 2. Molality is the most common concentration unit in chemistry.

______________ 3. Water is the most common solvent.

______________ 4. An amalgam is a solid metal-metal solution.

______________ 5. The solubility of a substance can be changed by altering the temperature.

______________ 6. In terms of physical states of matter, there are three possible combinations of solvent-solute pairs.

______________ 7. Gases have positive enthalpies of solution.

______________ 8. The differing solubilities of substances can be used to separate them from mixtures.

______________ 9. Solvation does not occur in the case of a polar solvent and a nonpolar solute.

______________ 10. The solution process is usually exothermic.

Answer the questions below.

11. Why does a solution of potassium chloride not exhibit the characteristic behavior of potassium chloride?

____

12. How is supersaturation possible?

____

____

____

13. What are three things that you could do to speed up the rate of solution?

____

14. What effect does pressure have on solutions in which the solute is a gas?

____

____

15. What is the difference between molarity and molality?

____

____

Name ______________________ Date ______________ Class ______________

16. In terms of polarity, which solvent-solute combinations are most likely to form solutions?

____

____

17. Complete the table shown as a review of the solubility rules for water solutions. For each substance listed, place a check mark in the correct column indicating whether you would predict it to be soluble or insoluble in water. (Use Appendix Table A-7 for reference.)

Substance	Soluble	Insoluble
$Al_2(SO_4)_3$		
$PbBr_2$		
$ZnCO_3$		
$CaBr_2$		
$(NH_4)_3PO_4$		
$Mn(NO_3)_2$		
$SrSO_4$		
MgF_2		
$Co(CH_3COO)_3$		

18. Match each substance a-c with the reagent(s) that can be used to precipitate the positive ion. Write your answer(s) on the lines provided.

Reagents:

a. $SrBr_2$ ______________ K_3PO_4

b. $MgCl_2$ ______________ LiI

c. $AgNO_3$ ______________ $Al_2(SO_4)_3$

Solve the following concentration problems.

19. If 10.0 g of sodium hydroxide is dissolved in 3.0 dm^3 of water, what is the molarity of the NaOH solution formed?

20. What is the molality of a solution formed by dissolving 60.0 g of HNO_3 in 2.50×10^3 g of water?

21. If 150 g of *n*-butylacetate, $C_6H_{12}O_2$, is dissolved in 190 g of ethanol, C_2H_6O, what is the mole fraction of each component of the solution?

Chapter 20

STUDY GUIDE

20.2 COLLOIDS

Match each type of colloid a through g with the correct example from the list that follows. Write the correct letter on the line provided. Some answers may be used more than once.

a. liquid emulsion
b. solid foam
c. solid sols
d. aerosols
e. solid emulsion
f. liquid foam
g. liquid sols

____ 1. mayonnaise
____ 2. whipped cream
____ 3. opals
____ 4. cheese
____ 5. jelly
____ 6. smoke
____ 7. pearls
____ 8. fog
____ 9. paint
____ 10. marshmallows

11. Complete the table comparing particles in solutions, suspensions, and colloids. Write *yes* or *no* for each characteristic to indicate whether it describes solutions, suspensions, or colloids.

Characteristic	Solutions	Suspensions	Colloids
less than 1 nm			
greater than 100 nm			
greater than 1 nm, but less than 100 nm			
settle out on standing			
pass unchanged through ordinary filter paper			
pass unchanged through membranes			
scatter light			
affect colligative properties			
particles can be seen with an ordinary light microscope			
heterogeneous			

Name ______________________ Date ______________ Class ______________

Answer each of the following questions on the lines provided.

12. What is colloid chemistry?

____

____

13. Why are colloids frequently used as the stationary phase in chromatography?

____

____

14. Why can semipermeable membranes be used to separate ions and colloidal particles?

____

Match each property of colloids with the letter of the correct description. Write your answer on the line provided.

a. random motion of colloidal particles
b. scattering of a beam of light
c. attraction and holding of substances on the surfaces of dispersed particles
d. migration of positive and negative colloidal particles in an electric field

____ 15. Tyndall effect
____ 16. adsorption
____ 17. Brownian motion
____ 18. electrophoresis

19. The Figures 1 and 2 illustrate two characteristic properties of colloids. In each case, identify the property and explain why it occurs.

Figure 1

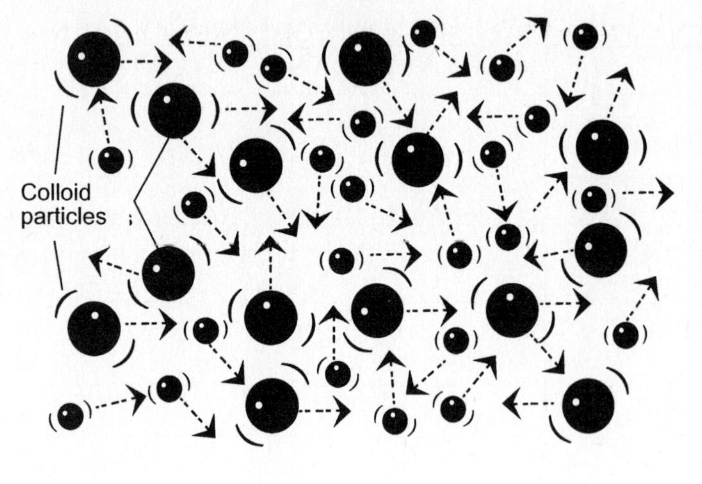

Figure 2

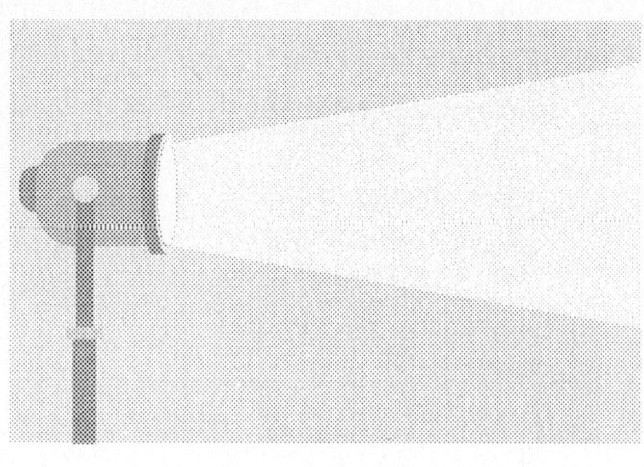

____

____

____

____

Name ______________________ Date ______________ Class ______________

Chapter 21

STUDY GUIDE

21.1 VAPOR PRESSURE CHANGES

For each of the following, write T *if the statement is true. If the statement is false, replace the underlined portion with a word or phrase that makes the statement true. Write your answer on the blank provided.*

______________ 1. Vapor pressure, freezing point, and boiling point are <u>chemical</u> properties of a solution.

______________ 2. Colligative properties depend on the <u>kind</u> of particles present in solution.

______________ 3. If fewer solvent particles can evaporate from a solution, the vapor pressure of the solvent is <u>lowered</u>.

______________ 4. For a solution having a specific concentration, the amount that the vapor pressure of the solvent is lowered depends on the characteristics of the <u>solute</u>.

______________ 5. Glucose, sucrose, and other molecular compounds with high melting points are considered <u>volatile</u> solutes.

______________ 6. The amount the vapor pressure of a solvent is lowered is equal to the difference between the vapor pressure of the pure solvent and the vapor pressure of the <u>solution</u>.

______________ 7. According to Raoult's law, the vapor pressure of a solution is directly proportional to the mole fraction of <u>solute</u>.

______________ 8. The vapor pressure of a sugar solution in which the mole fraction of sugar is 0.25 will be <u>4</u> times the vapor pressure of pure water.

______________ 9. In an ideal solution, all possible attractions among particles of solute and solvent are <u>the same</u>.

______________ 10. In an aqueous sucrose solution at 0°C, the mole fraction of water is 0.85. If the vapor pressure of water at 0°C is 0.6 kPa, its vapor pressure is lowered by <u>0.09</u> kPa.

Use the diagram of a fractional distillation apparatus shown here to answer the following questions about a solution consisting of pure liquids X and Y, where the boiling point of liquid X is greater than that of liquid Y.

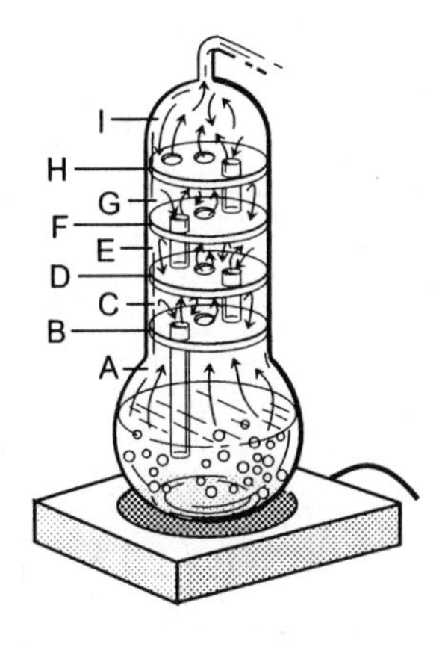

11. Which of the substances in the given solution is more volatile? Why?

____

12. When the vapor at A is in equilibrium with the given solution, is the vapor richer in substance X or substance Y?

____

13. The substances at D and E are in equilibrium. How does the temperature of the solution at D compare with the temperature of the vapor at E?

____

14. How does the temperature at I compare with the temperature at A? Explain your reasoning.

____

15. How does the composition of the substances at the top of the apparatus (H and I) differ from the composition of the substances at B and C?

____

Use Table 21.1 in your textbook to answer the following questions about the fractional distillation of petroleum.

16. Which fraction(s) would come off first? Explain your reasoning.

____

17. Which fraction(s) would collect at the bottom of the fractionating tower?

____

18. Would the fractions at the middle levels of the tower be richer in fuel oil or jet fuel?

____

Use the symbol > (greater than) or < (less than) to complete each of the following comparisons between vapor pressures, boiling points, and freezing points of (1) a pure solvent, and (2) a solution of a nonvolatile solute in the solvent at the same temperature.

19. vapor pressure: pure solvent ____ solution
20. boiling point: pure solvent ____ solution
21. freezing point: pure solvent ____ solution

Answer each of the following questions.

22. Describe and explain the effect of adding a nonvolatile, nonionizing solute on the boiling point of a pure solvent.

____

____

____

23. Why is the addition of antifreeze to the water in a car radiator just as important in hot climates as in cold climates?

____

____

Name ______________________ Date ______________ Class ______________

Chapter 21

STUDY GUIDE

21.2 QUANTITATIVE CHANGES

Answer each of the following questions.

1. What is meant by the molal boiling point constant and molal freezing point constant for a given solvent?

____

____

____

2. Describe freezing point depression in terms of the freezing points of a pure solvent and a solution made from that solvent.

____

____

3. Ideally, a 1*m* solution of zinc chloride ($ZnCl_2$) in water would boil at 101.545°C, whereas a 1*m* solution of glucose ($C_6H_{12}O_6$) would boil at 100.515°C.

   a. What is the boiling point elevation of an ideal solution of zinc chloride in water?

   ____

   b. Compare the boiling point elevation of a zinc chloride solution with the boiling point elevation of a glucose solution.

   ____

   c. How does the difference in boiling point elevation provide evidence that zinc chloride dissociates in water?

   ____

   ____

   ____

4. The molal freezing point constant for pure water is 1.853 C°/*m*. Assuming 100% dissociation, at what temperature would a 1*m* solution of magnesium chloride ($MgCl_2$) in water freeze?

____

5. A solution contains 20.0 g maltose ($C_{12}H_{22}O_{11}$), a nonvolatile, nonionizing solute, dissolved in 324 g H_2O. Fill in the missing parts and complete the following calculations for the boiling and freezing points of the resulting solution.

   a.
$20.0\ g\ C_{12}H_{22}O_{11}$	$1\ mol\ C_{12}H_{22}O_{11}$	
$324\ g\ H_2O$		$1\ kg\ H_2O$

   = ______________________

   b. boiling point elevation = (______________________)(0.515 C°/~~*m*~~) = ______________________
   boiling point = ______________________ °C

   c. freezing point depression = (______________________) (1.853 C°/~~*m*~~) ______________________

   freezing point = ______________________ °C

In the diagrams shown, a semipermeable membrane covers the opening of the test tube that is inverted in the beaker. The molecules of solvent used can pass through the membrane, but the molecules of solute cannot. Study the diagrams, then answer the questions that follow.

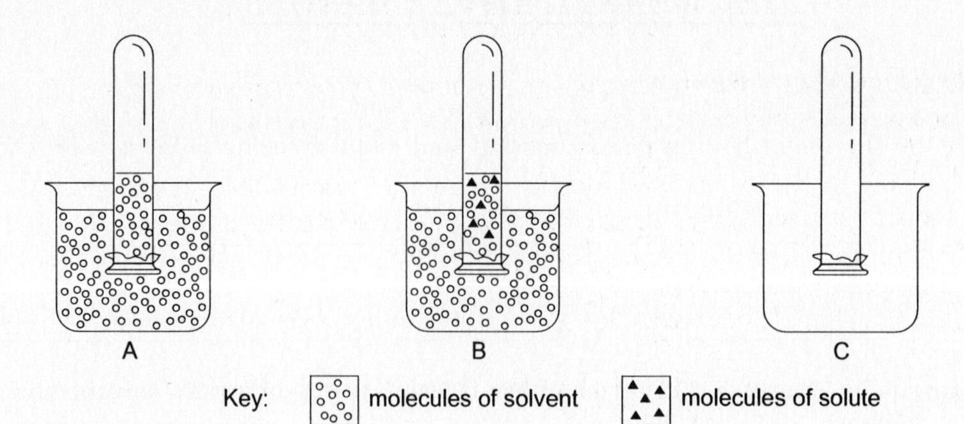

6. Draw arrows on diagram A to indicate the direction and rate of movement of solvent molecules. Explain the reasoning for what you drew.

____

7. Describe the direction and rate of movement of solvent and solute molecules in diagram B. Draw arrows to represent the movement of solvent molecules.

____

____

8. In terms of solvent concentration, describe the net movement of molecules in diagram B.

____

9. On which side of the membrane in diagram B will an increase in volume take place?

____

10. At equilibrium, on which side of the membrane will osmotic pressure be exerted?

____

11. On diagram C, show the solution and the pure solvent at equilibrium. Use arrows to indicate the flow of solvent molecules.

For each of the following, write T *if the statement is true. If the statement is false, replace the underlined portion with a word or phrase that makes the statement true. Write your answer on the line provided.*

______________ 12. Ion activity is an indication of the degree to which an ionic compound <u>actually</u> dissociates in a water solution.

______________ 13. In reality, ionizing ionic compounds in aqueous solution do not completely dissociate because of the <u>repulsive</u> forces between <u>like</u> charged ions.

______________ 14. For an ionizing ionic compound dissolved in water, the more dilute the solution, the <u>less</u> effective each ion is in freezing point depression and boiling point elevation.

Chapter 22

STUDY GUIDE

22.1 REACTION RATES

Write T for true or F for false. If the statement is false, replace the underlined word or phrase with one that makes the statement true and write your correction on the blank provided.

______ 1. A substance that does not break down spontaneously at room temperature is said to be kinetically unstable.

______ 2. A reaction that eventually reaches equilibrium is said to be reversible.

______ 3. A reaction goes to completion if at least one of the reactants is used up.

______ 4. The rate of a reaction from left to right is the rate at which a product disappears.

______ 5. Heat, a flame, or a spark may supply the activated complexes needed for a reaction to take place.

______ 6. The reactions of kinetically stable substances tend to have high activation energies.

______ 7. The energy required to form an activated complex is converted to kinetic energy.

______ 8. Ionic reactions tend to have faster reaction rates than electron transfer reactions.

______ 9. Reaction rate varies directly as the product of the masses of the reactants.

______ 10. Specific rate constant refers to the rate at which a reaction occurs at a given temperature.

______ 11. For a given reaction, the total number of molecules having the required activation energy at 125°C would be the same as the total number of molecules having the required activation energy at 100°C.

______ 12. A substance that affects reaction rate without being chemically changed at the end of the reaction is an inhibitor.

Match each situation in questions 13-18 with the letter of the term that best describes it. A letter may be used more than once.

a. homogeneous catalyst
b. homogeneous reaction
c. heterogeneous catalyst
d. heterogeneous reaction
e. inhibitor
f. kinetically stable
g. thermodynamically stable

___ 13. $H_2O(g) + CO(g) \rightarrow H_2(g) + CO_2(g)$

___ 14. The reaction between $N_2(g)$ and $O_2(g)$ does not take place at room temperature because high activation energies are required.

___ 15. $CaO(cr) + CO_2(g) \rightarrow CaCO_3(cr)$

___ 16. Manganese dioxide powder is added to increase the decomposition rate of liquid potassium chlorate.

___ 17. BHT is added to food as a preservative.

___ 18. $C_2H_5OH(aq) + CH_3COOH(aq) \rightleftarrows CH_3COOC_2H_5(aq) + H_2O(l)$

Name ______________________ Date ______________________ Class ______________________

For each of the following reactions, write I *if the effect of the given change is an increase in reaction rate,* D *if the effect is a decrease in reaction rate, or* RS *if the reaction rate remains the same. Assume that the reaction takes place in a closed vessel of fixed volume.*

19. Limestone (calcium carbonate) reacts with hydrochloric acid in an irreversible reaction, to form carbon dioxide and water as described by the following equation:
$CaCO_3(cr) + 2HCl(aq) \rightarrow CaCl_2(aq) + CO_2(g) + H_2O(l)$
What is the effect if

______ a. the temperature is lowered?

______ b. the volume of the reaction vessel is increased?

______ c. limestone chips are used instead of a block of limestone?

______ d. the pressure inside the reaction vessel is increased?

______ e. a more dilute solution of HCl is used?

20. At a given temperature, ethene, C_2H_4, reacts with chlorine in a reversible reaction, to produce 1,2-dichloroethane ($C_2H_4Cl_2$) as described by the following equation:
$C_2H_4(g) + Cl_2(g) \rightarrow C_2H_4Cl_2(g)$
What is the effect if

___ a. the volume of the reaction vessel is reduced?

___ b. the pressure inside the reaction vessel is reduced?

___ c. a catalyst is added?

___ d. the volume of the reaction vessel is doubled and the pressure inside the vessel is halved?

Use the energy diagram shown to answer questions 21 - 29.

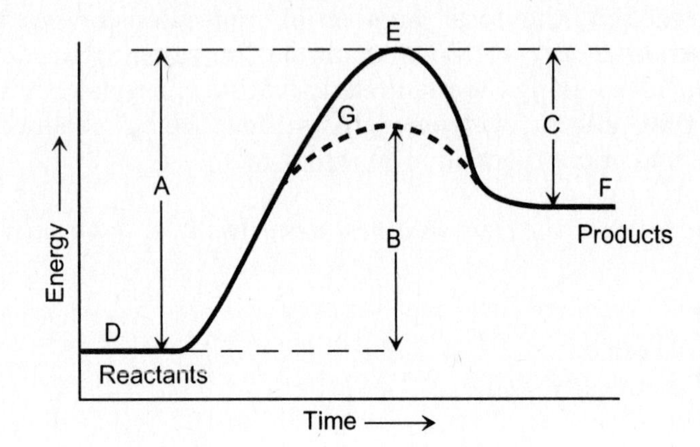

___ 21. Which letter indicates the energy content of the reactants?

___ 22. Which letter indicates the energy content of the products?

___ 23. Which letter indicates the change in energy required for the formation of activated complexes in the forward reaction?

___ 24. Assume that the reaction is reversible. Which line indicates the change in energy required for the formation of activated complexes in the reverse reaction?

___ 25. Which letter represents the energy of the activated complexes in the uncatalyzed reaction?

___ 26. Which letter indicates the reaction in the presence of a catalyst?

___ 27. Which letter indicates the activation energy needed for the forward reaction if a catalyst is used?

28. How are the activation energies of the forward and reverse reactions affected by the addition of a catalyst?

______

29. How does the potential energy of the activated complexes of the forward reaction compare with the potential energy of the activated complexes of the reverse reaction?

______

30. For the reaction A + B → C, the following data were obtained:

Trial	[A]	[B]	Rate
1	0.10*M*	0.20*M*	0.00003 $(mol/dm^3)/s$
2	0.10*M*	0.40*M*	0.00006 $(mol/dm^3)/s$
3	0.20*M*	0.40*M*	0.00048 $(mol/dm^3)/s$

a. How is the rate of reaction affected by a change in concentration of reactant A? Reactant B?

______

b. If the rate expression for this reaction is: rate = $k[A]^3[B]$, find k.

Name ______________________ Date ______________ Class ______________

Chapter 22

STUDY GUIDE

22.2 CHEMICAL EQUILIBRIUM

For items 1-8, underline the term inside the parentheses that makes each statement true.

1. At equilibrium, the rate of the forward reaction is (equal to, greater than) the rate of the reverse reaction.

2. The equilibrium constant for a given reaction at a given temperature is the (product, quotient) of the specific rate constant for the forward reaction and the specific rate constant for the reverse reaction.

3. The exponents used in the expression for the equilibrium constant are the (subscripts, coefficients) of the reactants and products.

4. If a reaction tends to go toward completion, the rate of the forward reaction is (equal to, greater than) the rate of the reverse reaction before equilibrium is reached.

5. If $K_{eq} = 1.2 \times 10^{-5}$, the concentration of the reactants is (greater than, less than) the concentration of the products at equilibrium.

6. At temperature T_1, K_{eq} for a certain reaction is 0.239. At temperature T_2, K_{eq} for the same reaction is 4.7. By changing the temperature from T_1 to T_2, the equilibrium will shift in favor of the (reactants, products).

7. If the reaction $H_2(g) + Cl_2(g) \rightleftarrows 2HCl(g)$ + *heat* is at equilibrium, a decrease in (volume, temperature) will produce a shift in equilibrium toward the right.

8. An increase in pressure on the system $2CO_2(g) \rightleftarrows 2CO(g) + O_2(g)$ at equilibrium results in an equilibrium shift toward the (left, right).

Answer or complete each of the following items.

9. What factors can affect the equilibrium of a reaction?

____

10. A reversible one-step reaction occurs between carbon monoxide gas, CO, and hydrogen gas, H_2, to produce methane gas, CH_4, and gaseous water. Using this information, fill in the diagram according to the following guidelines:
    a. Within the ovals, write the balanced equation for the reaction.
    b. Label the forward and reverse reactions on the lines provided.
    c. In the rectangles, write the words *reactants* or *products*, as appropriate, to represent both the forward and reverse reactions.

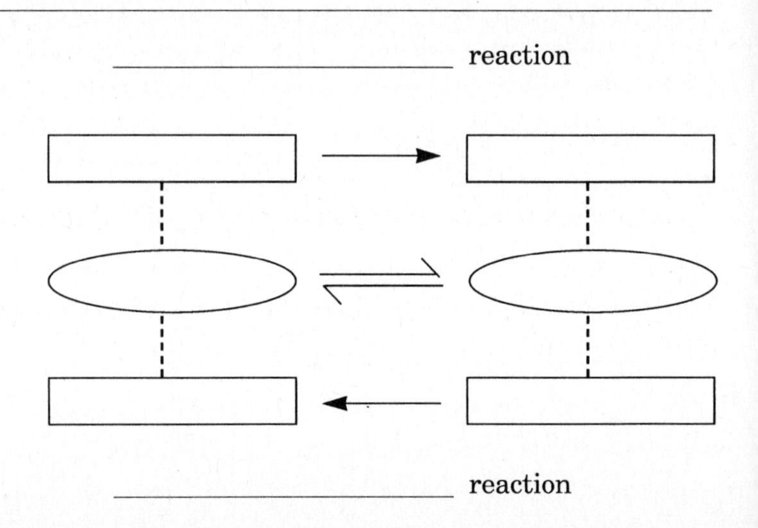

11. Write the expression for the equilibrium constant for the reaction in question 10.

12. If temperature and pressure of the reaction in question 10 are kept constant, but the concentration of each of the substances is halved,
a. how does the rate of the forward reaction change?

______

b. how does the rate of the reverse reaction change?

______

c. what would be the net change in the relative amounts of reactants and products?

______

13. In the reaction in question 10, if the volume of the reaction vessel and temperature are kept constant and the pressure on the system is increased,

a. the concentrations of which substances would be affected? ______

b. in which direction will the equilibrium shift? ______

c. which substance(s) will show an increase in concentration when equilibrium is reestablished?

______

14. Consider the equilibrium equation for the reaction: $4HCl(g) + O_2(g) + heat \rightleftarrows 2Cl_2(g) + 2H_2O(g)$

a. If the temperature is increased, the reaction will favor the formation of

______.

b. Which reaction requires an input of energy? ______

15. In the Haber process, which involves the reaction $N_2(g) + 3H_2(g) \rightleftarrows 2NH_3$ + energy,

a. why is NH_3 removed as it is formed?

______

b. why is the use of a catalyst considered one of the optimum conditions for this process?

______

c. what is the effect on the relative amounts of product and reactant if the catalyst is removed?

______

16. Given $K_{eq} = \frac{[NO]^4[H_2O]^6}{[NH_3]^4[O_2]^5}$

a. Write the chemical equation for the reversible reaction having the given K_{eq}.

b. At a certain temperature the concentrations of NO and NH_3 are equal, and the concentration of H_2O and O_2 are 2.0*M* and 3.0*M* respectively. What is the value of K_{eq} at this temperature?

Chapter 23

STUDY GUIDE

23.1 ACIDS AND BASES

Complete the sentence or answer the question.

1. Define or explain the following theories about acids and bases.
   a. Arrhenius Theory ______
   b. Brønsted-Lowry Theory ______
2. A(n) ______ is an acid, base, or salt, which, when dissolved in water, conducts an electric current.
3. The polyatomic ion, H_3O^+, which is formed when a hydrogen ion (H^+) combines with a water molecule, is called the ______.
4. The ______ of an acid is the particle that remains after a proton has been released by the acid.
5. The ______ of a base is formed when the base acquires a proton from the acid.
6. Complete the following equation that, *in general,* describes any acid-base reaction:

   ______ + ______ → ______ + ______
7. Acids that contain only two elements are called ______.
8. Acids that contain three elements are called ______.
9. Substances that can react as either an acid or a base are said to be ______.
10. The most common example of the kind of compound described in exercise 9 is ______.

Match each term with the correct description.

a. acidic anhydride
b. basic anhydride
c. weak acid
d. strong acid
e. strong base
f. weak base

___ 11. a base that is completely dissociated in solution into positive ions and negative ions
___ 12. any oxide that will produce an acid when dissolved in water
___ 13. an acid that is considered to ionize completely in solution into positive ions and negative ions
___ 14. any oxide that will produce a base when dissolved in water
___ 15. slightly ionized acid in solution
___ 16. a base that is only partially ionized in solution

Name ______________________ Date ______________________ Class ______________________

Complete the following exercises.

17. Name the following binary acids.

    a. HF ______________________
    b. HCl ______________________
    c. HI ______________________
    d. H_2S ______________________
    e. HN_3 ______________________
    f. HBr ______________________

18. Name the following ternary acids and bases.

    a. H_3AsO_4 ______________________
    b. H_3AsO_3 ______________________
    c. NaOH ______________________
    d. H_3BO_3 ______________________
    e. H_3PO_4 ______________________
    f. H_3PO_3 ______________________
    g. HNO_3 ______________________

19. Name the following organic acids and bases.

    a. CH_3NH_2 ______________________
    b. CH_3COOH ______________________
    c. CH_3CH_2COOH ______________________

20. Write the formula for the conjugate base of each of the acids listed.

    a. H_2O ______________________
    b. HNO_3 ______________________
    c. HF ______________________
    d. $HC_2H_3O_2$ ______________________

21. Write the formula for the conjugate acid of each of the following bases listed.

    a. NH_3 ______________________
    b. HSO_4^- ______________________
    c. HS^- ______________________
    d. C_2H_5NH ______________________

22. Classify each of the following as a strong acid, strong base, weak acid, or weak base.

    a. NaOH ______

    b. HCl ______

    c. NH_4^+ ______

    d. NH_3 ______

    e. Cl^- ______

    f. HI ______

23. In the following reaction, which species behave as Brønsted acids? As Brønsted bases?
    $H_2SO_4(aq) + H_2O \rightleftarrows HSO_4^-(aq) + H_3O^+(aq)$

    ______

    ______

24. Write the formulas for the anhydrides of the following.

    a. H_2CO_3 ______

    b. H_2SO_3 ______

    c. H_2SO_4 ______

    d. NaOH ______

    e. $Ca(OH)_2$ ______

    f. HNO_3 ______

Name ______________________ Date ______________ Class ______________

Chapter 23

STUDY GUIDE

23.2 SALTS AND SOLUTIONS

Complete the sentence or answer the question.

1. Define or explain the following terms.

   a. salt ____

   ____

   b. neutralization reaction ____

   ____

   c. polyprotic acid ____

   ____

2. Ions present in a solution but not involved in the reaction are called ______________.

3. State the general rule that must be observed when writing net ionic equations.

   ____

   ____

4. Convert the following balanced equations first to ionic form and then to net ionic form.

   a. $BaCl_2 + Na_2SO_4 \rightarrow BaSO_4 + 2NaCl$

   b. $Na_2CO_3 + 2HCl \rightarrow 2NaCl + CO_2 + H_2O$

   c. $NiS + 2HCl \rightarrow NiCl_2 + H_2S$

5. Select the polyprotic acids from the following list.

   HCN, H_2CO_3, HCl, HF, H_2SO_3, H_3PO_4, H_2SeO_3, HBr

   ____

Write T *for true or* F *for false. If a statement is false, replace the underlined word or phrase with one that will make the statement true, and write the correction on the blank provided.*

______________ 6. A salt is formed as a product in the equilibrium reaction of an acid and a base.

______________ 7. Salts may be formed from the reaction of acidic or basic anhydrides with a corresponding acid, base, or anhydride.

______________ 8. The percent ionization of a weak acid or base dissolved in water is a measure of the amount of ionization of the acid or base in solution.

Answer each question.

9. Write the equilibrium constant expression, K_{eq}, for the following acid ionization reaction.
   $HOBr + H_2O \rightleftarrows H_3O^+ + OBr^-$

10. Write the ionization constant expression, K_a, for the reaction in question 9.

11. Calculate the hydronium ion concentration of a 0.25*M* solution of hypochlorous acid, HOCl, for which $K_a = 3.5 \times 10^{-8}$. The equation for ionization of this acid is $HOCl + H_2O \rightleftarrows H_3O^+ + OCl^-$.

12. Calculate the percent of ionization of 0.010*M* acetic acid solution, if the hydronium ion concentration is $4.2 \times 10^{-4} M$.

Chapter 24

STUDY GUIDE

24.1 WATER EQUILIBRIA

Complete the sentence or answer the question.

1. Define or explain the following. (Include formulation where applicable.)
   a. solubility product constant, K_{sp} of a saturated solution ______

   b. ion product constant of water, K_w ______

   c. pH ______

   d. hydrolysis ______

   e. buffer solution ______

2. The addition of a common ion to a saturated solution ______ the solubility of a substance in solution.

3. Write the solubility product expression for each of the following.
   a. $BaSO_4$ ______
   b. CaF_2 ______
   c. Bi_2S_3 ______
   d. $MgNH_4PO_4$ ______
   e. Ag_2CrO_4 ______

4. The concentration of water in pure water is ______ mol/dm^3.

5. Write the formula for calculating the pH from the concentration of H_3O^+.

______

6. Find the pH of solutions with the following H_3O^+ concentrations.
   a. $4.21 \times 10^{-4} M$

   b. $5.00 \times 10^{-6} M$

   c. $9.11 \times 10^{-8} M$

   d. $2.14 \times 10^{-13} M$

   e. $5.21 \times 10^{-10} M$

   f. $1.45 \times 10^{-2} M$

Match the acid-base reaction type with the examples given.

a. weak base + strong acid
b. strong acid + strong base
c. weak acid + strong base

___ 7. $H^+ + OH^- \rightleftarrows H_2O$
___ 8. $HC_2H_3O_2 + OH^- \rightleftarrows H_2O + C_2H_3O_2^-$
___ 9. $NH_3 + H^+ \rightleftarrows NH_4^+$

10. Using the rules outlined in Chapter 24, classify each of the following solutions as acidic, basic, or neutral.

a. $NaNO_3$ ______________
b. KCN ______________
c. Na_2CO_3 ______________
d. NH_4Br ______________
e. $AlCl_3$ ______________

Write T for true or F for false. If a statement is false, replace the underlined word or phrase with one that will make the statement true, and write your correction on the blank provided.

______________ 11. Buffers are <u>most</u> efficient at neutralizing added acids or bases when the concentrations of HA and A^- (or MOH and M^+) are the same.

______________ 12. The pH is the measure of <u>oxygen</u> in a solution.

______________ 13. Pure water is a <u>good</u> conductor of electricity.

______________ 14. <u>K_w</u> is a constant for all dilute aqueous solutions at room temperature.

______________ 15. As the hydronium concentration increases in a solution, the solution is made more <u>basic</u>.

Answer each question.

16. Label the parts of the pH scale in terms of acid, base, and neutral segments.

pH = 0 ◄———————— pH = 7 ————————► pH = 14

17. What is the concentration of silver in a saturated solution of $AgC_2H_3O_2$? The K_{sp} of $AgC_2H_3O_2$ is 2.00×10^{-3}.

18. Will a precipitate of $CaSO_4$ form in **hard water** if the Ca^{2+} concentration is 0.010*M* and the SO_4^{2-} concentration is 0.00100*M*? Show your calculations. (K_{sp} of $CaSO_4 = 3.0 \times 10^{-5}$)

19. If 0.40 dm^3 of 0.020*M* Na_2SO_4 is added to 0.80 dm^3 of hard water, where $[Ca^{2+}]$ = 0.010*M* (as in question 18), will a precipitate form? Show your calculations.

Chapter 24

STUDY GUIDE

24.2 TITRATION

Complete the sentence or answer the question.

1. Define or explain the following.

   a. indicator ______________________________

   ____

   b. titration ______________________________

   ____

   c. standard solution ______________________________

2. Explain what a pH meter is and what it is used for.

   ____

   ____

3. Explain how a titration is carried out.

   ____

   ____

   ____

   ____

4. A titration is performed in which 22.3 cm^3 of 0.240*M* NaOH reacts with a 50.0-cm^3 sample of $HC_2H_3O_2$. What is the concentration of the $HC_2H_3O_2$?

Write T *for true or* F *for false. If a statement is false, replace the underlined word or phrase with one that will make the statement true, and write your correction on the blank provided.*

______________ 5. Titration indicators are most useful when they are used on <u>colorful</u> solutions.

______________ 6. If an acid is added to a base, a <u>neutralization reaction</u> occurs.

______________ 7. Many indicators must be used in order to test for pH over a <u>wide</u> range of the pH scale.

Chapter 25

STUDY GUIDE

25.1 OXIDATION AND REDUCTION PROCESSES

Complete each of the following.

1. A reaction in which ions or atoms undergo changes in electron structure is referred to as a(n) ______ reaction.
2. The loss of electrons from an atom or ion is called ______.
3. The gain of electrons by an atom or ion is called ______.
4. The substance in a redox reaction that undergoes oxidation is referred to as a(n) ______ agent.
5. The substance in a redox reaction that undergoes reduction is referred to as a(n) ______ agent.
6. In the reaction that follows, ______ is oxidized and ______ is reduced. $H_2 + Cl_2 \rightarrow 2HCl$
7. The oxidation number of a free element is ______.
8. In most reactions, the oxidation number of the calcium ion is ______, while that for the oxide ion is generally ______.
9. The sum of the oxidation numbers of all of the atoms in HNO_3 is ______.
10. The oxidation number of N in $Mg(NO_3)_2$ is ______.
11. The oxidation number of S in SCl_2 is ______.
12. In the reaction that follows, the oxidizing agent is ______. $C + H_2O \rightarrow CO + H_2$
13. In the reaction given in question 12, the element being oxidized is ______.
14. In the reaction given in question 12, the total number of electrons transferred (lost or gained) is ______.
15. The oxidation number of P in PO_4^{3-} is ______.
16. Write true (T) or false (F) for each of the following.

____ a. The following is an example of a redox reaction: $HCl + NaOH \rightarrow NaCl + H_2O$

____ b. The term *oxidation* was first applied to the combining of oxygen with other elements.

____ c. In an oxidation-reduction reaction, the number of electrons lost must equal the number gained.

____ d. If a substance gains electrons readily, it is said to be a strong reducing agent.

____ e. The oxidation number of a monatomic ion is equal to the charge on the ion.

____ f. In general, ions of the Group 17 elements have oxidation numbers of 1–.

____ g. The oxidizing agent in the following reaction is Na. $2Na + S \rightarrow Na_2S$

____ h. A total of 2 electrons are transferred in the reaction given in question 16g.

____ i. The element that undergoes reduction in the reaction given in question 16g is S.

Name ______________________ Date ______________ Class ______________

17. Complete the table below for each reaction specified.

Reaction	Element oxidized	Element reduced	Oxidizing agent	Reducing agent	Total e^- transferred
$2HBr + Cl_2 \rightarrow 2HCl + Br_2$					
$Zn + I_2 \rightarrow ZnI_2$					
$Fe_2O_3 + 3CO \rightarrow 2Fe + 3CO_2$					
$16H^+ + 2MnO_4^- + 5C_2O_4^{2-} \rightarrow 2Mn^{2+} + 8H_2O + 10CO_2$					

Chapter 25

STUDY GUIDE

25.2 BALANCING REDOX EQUATIONS

1. Complete the steps outlined below in order to balance the following oxidation-reduction reaction: $HNO_3 + S \rightarrow NO_2 + H_2SO_4 + H_2O$

   a. Write the skeleton equation for the reaction.

   b. Assign oxidation numbers to all atoms involved in the reaction.

   c. Identify the substance being oxidized, and then write and balance the oxidation reaction in terms of both atoms and charge. In acidic solutions, it may be necessary to add H^+ ions and H_2O molecules to balance the reaction. In basic solutions, it may be necessary to add a sufficient number of OH^- ions to each side of the equation to combine with any excess H^+ ions and form H_2O molecules.

   d. Identify the substance being reduced, and then write and balance the reduction reaction in terms of both atoms and charge.

   e. Combine the two half-reactions, first multiplying one or both by the factor(s) needed to balance the electron transfer. Cancel the electrons, as well as the excess number of any species that appear on both sides of the equation, from the final reaction.

   f. Perform a final check on the balanced equation to ensure that both atoms and charge are balanced.

Apply the steps given above to balance each of the following oxidation-reduction reactions.

2. $Br_2 + SO_2 + H_2O \rightarrow H_2SO_4 + HBr$

3. $PbS + H_2O_2 \rightarrow PbSO_4 + H_2O$

4. $H_3AsO_4 + Zn \rightarrow AsH_3 + Zn^{2+}$

5. $PH_3 + O_2 \rightarrow P_4O_{10} + H_2O$

6. $NO_2 + OH^- \rightarrow NO_2^- + NO_3^-$ (in basic solution)

7. $K_2Cr_2O_7 + Na_2SO_3 + HCl \rightarrow Cr_2(SO_4)_3 + KCl + NaCl + H_2O$

8. $Cu(OH)_2 + HPO_3^{2-} \rightarrow Cu_2O + PO_4^{3-}$

9. $TeO_2 + BrO_3^- \rightarrow H_6TeO_6 + Br_2$

10. $Bi_2S_3 + NO_3^- \rightarrow Bi^{3+} + NO + S$

11. $NO + H_5IO_6 \rightarrow NO_3^- + IO_3^-$

Chapter 26

STUDY GUIDE

26.1 CELLS

1. Indicate if the bulb shown in Figure 1 will light or not by placing a plus(+) or minus(–) sign, respectively, in the blank to the left of each of the following situations.

____ a. The electrodes are connected with a piece of glass.

____ b. The beaker is filled with air.

____ c. The beaker is filled with an electrolyte.

____ d. The electrodes are connected by a wire.

____ e. The beaker is filled with alcohol.

Figure 1

2. In situation c of question 1, write the type of conduction, if any, of the materials.

a. wires ______ c. bulb filament ______

b. electrolyte ______ d. electrodes ______

Write T for true or F for false. If a statement is false, replace the underlined word or phrase with one that will make the statement true, and write your correction on the blank provided.

______ 3. A galvanometer is an instrument that detects electric current.

______ 4. Potential difference is measured in units of amperes.

______ 5. Current flows through wires in which there is no potential difference between the ends.

______ 6. Metallic conduction takes place because the electrons of metals are free to move when a small potential difference is applied.

______ 7. Electrolytic conduction takes place because electrons move freely in aqueous solutions.

Complete the sentence.

8. An ______ is any substance that produces ions in solution.

9. The process by which an electric current produces a chemical change is called ______.

10. In a cell, oxidation and reduction occur as separate half-reactions at the same ______.

11. A device that converts chemical potential energy into electric energy is called a ______.

12. The zinc-copper cell reaction, in which Zn is oxidized to Zn^{2+} and Cu^{2+} is reduced to Cu, is represented as ______.

Name ______________ Date ______________ Class ______________

Using the letters of the terms below, label each of the diagrams in Figure 2.

a. anode (Pb)
b. anode (Zn can)
c. anode (Zn shell)
d. automobile battery
e. cathode (graphite)
f. cathode (PbO_2)
g. cathode (steel)
h. dry cell
i. electrolyte (H_2SO_4)
j. electrolyte (KOH and paste of $Zn(OH)_2$ and HgO)
k. electrolyte (paste of MnO_2, NH_4Cl, and powdered graphite)
l. mercury battery

Figure 2

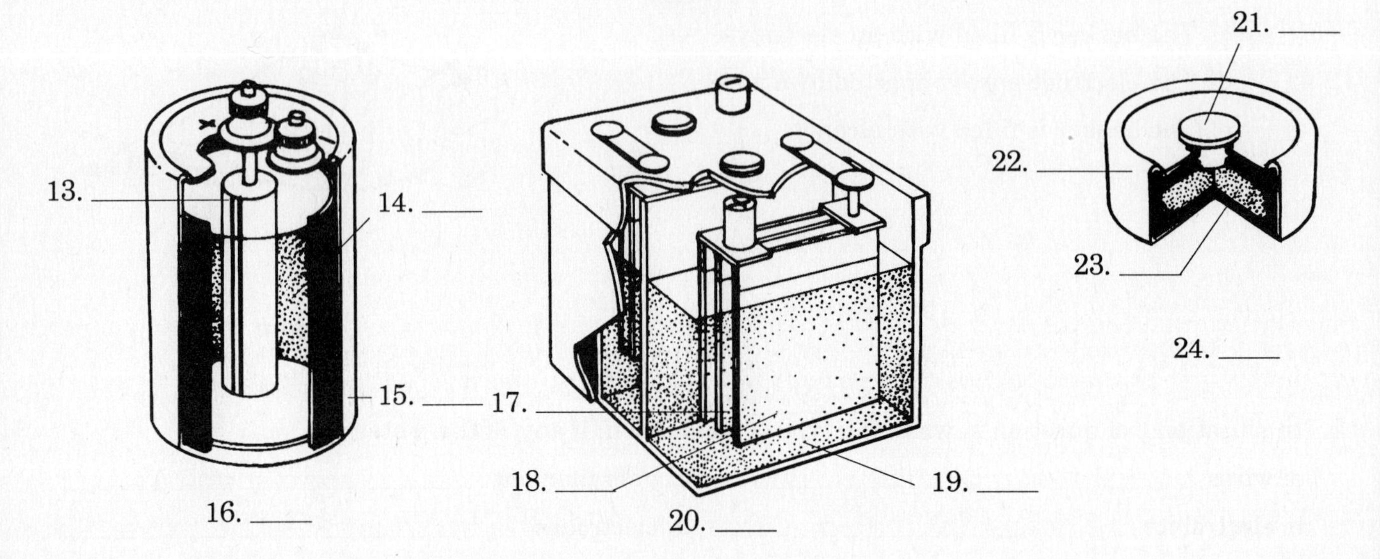

Chapter 26

STUDY GUIDE

26.2 QUANTITATIVE ELECTROCHEMISTRY

Write T *for true or* F *for false. If a statement is false, replace the underlined word or phrase with one that will make the statement true, and write your correction on the blank provided.*

______________ 1. The difference between two half-cells is a measure of the relative tendency of the two substances to take on protons.

______________ 2. The half-reactions listed in Table 26.1 of the text are written as oxidations.

______________ 3. The symbol for standard reduction potential (25°C, 101.325 kPa, 1*M*) is E^0.

______________ 4. The half-reaction of carbon is assigned a reduction potential of 0.000 0 V.

______________ 5. The standard reduction potential of a substance is an intensive property.

Use Table 26.1 in the text to answer questions 6-11.

6. Rank the following substances in order of decreasing reduction strength.
   Ag^+ Au^+ Cs^+ K^+ Li^+ Na^+

   ____

7. Explain why the reaction, $Cu(cr) + 2Ag^+(aq) \rightarrow 2Ag(cr) + Cu^{2+}(aq)$ will occur spontaneously.

   ____

   ____

   ____

8. Predict if each of the following reactions will occur spontaneously as written.

   a. $Cu(cr) + Pd^{2+}(aq) \rightarrow Cu^{2+}(aq) + Pd(cr)$

   b. $2Ag(cr) + Co^{2+}(aq) \rightarrow Co(cr) + 2Ag^+(aq)$

9. Consider the following cell (standard conditions).

$Zn|Zn^{2+}||Ni^{2+}|Ni$

a. What substance is oxidized in the cell?

b. Write the oxidation half-cell reaction.

c. What is the potential for the oxidation half-cell reaction?

d. What substance is reduced in the cell?

e. What is the reduction half-cell reaction?

f. What is the potential for the reduction half-cell reaction?

g. What is the potential produced by the cell?

10. Predict the potential for each of the following cells.

a. $Zn|Zn^{2+}||Cu^{2+}|Cu$

b. $Fe|Fe^{2+}||Ag^{+}|Ag$

11. The relationship $E = -0.05916$ pH indicates that E is a ________________ function of pH.

12. The following cell has a standard potential of 0.459V.

$Cu|Cu^{2+}||Ag^{+}|Ag$

Write the chemical reaction for the cell.

13. In the spaces below, write the values that will complete the expression for the potential of the following cell.
$Cu|Cu^{2+}(2.0M)||Ag^{+}(1.5M)|Ag$

$$(a) - \frac{0.05916}{(b)} \log \frac{[(c)]^{(e)}}{[(d)]^{(f)}}$$

a. ______________________

b. ______________________

c. ______________________

d. ______________________

e. ______________________

f. ______________________

14. Predict whether the potential of the cell in question 13 will be greater or less than that of the standard cell in question 12.

____

15. In the spaces below, write the values that will complete the expression for calculating the mass of copper deposited by a current of 6.0 amperes flowing for a period of 1.00 min.

(a)	1 min	(b)	1C	1 mol e^-	1 mol Cu	(d) Cu
		1 min	A · s	96 485 C	(c) mol e^-	1 mol Cu

a. ______________________

b. ______________________

c. ______________________

d. ______________________

Chapter 27

STUDY GUIDE

27.1 INTRODUCTORY THERMODYNAMICS

1. For each of the following processes, determine whether the indicated quantities of the system are greater than zero, equal to zero, or less than zero. Write your answer in the blank following each quantity. If not enough information is given to determine a value, place an X in the blank.

   a. A system loses heat.
   q _____ w _____ ΔU _____

   b. A sample of a gas is compressed without heat transfer.
   q _____ w _____ ΔU _____

   c. A system does work without a decrease in its internal energy.
   q _____ w _____ ΔU _____

   d. A sample of a gas is heated. As the gas expands, it does work pushing a piston.
   q _____ w _____ ΔU _____

   e. The internal energy of a confined gas increases without work being done.
   q _____ w _____ ΔU _____

Figure 1 is a pressure-temperature graph of a system showing two paths (arrows) that represent independent ways of changing the system. The internal energy of the system is given by $\Delta U = q + w$. Use Figure 1 to answer questions 2-9.

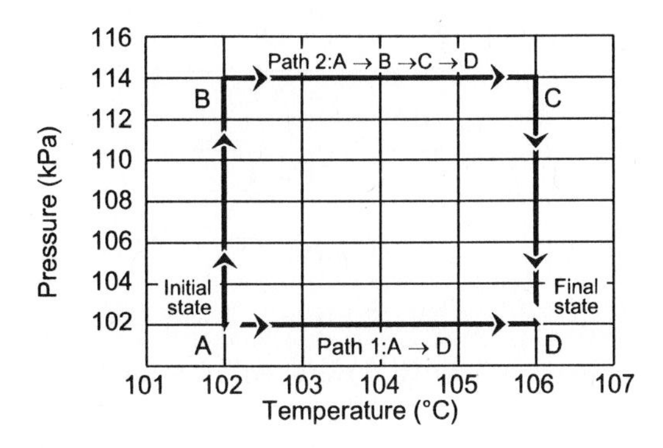

Figure 1

2. Underline each of the following variables that is a state function.

   P T q w ΔU

3. Underline each of the following paths that represents an isobaric process.

   A→B B→C C→D A→D

4. Underline each of the following paths that represents an isothermal process.

   A→B B→C C→D A→D

5. What is the final temperature of the system?

____

Name ______________________ Date ______________ Class ______________

6. What is the final pressure of the system?

7. What is ΔT of the system?

8. What is ΔP of the system?

9. If ΔH of the system is 20.0 J, determine q_p

Figure 2 represents the changes of enthalpy of the reactions

$H_2(g) + \frac{1}{2}O_2(g) \rightarrow H_2O(l)$

$H_2O(l) \rightarrow H_2O(g)$

Use Figure 2 to answer questions 10 - 14.

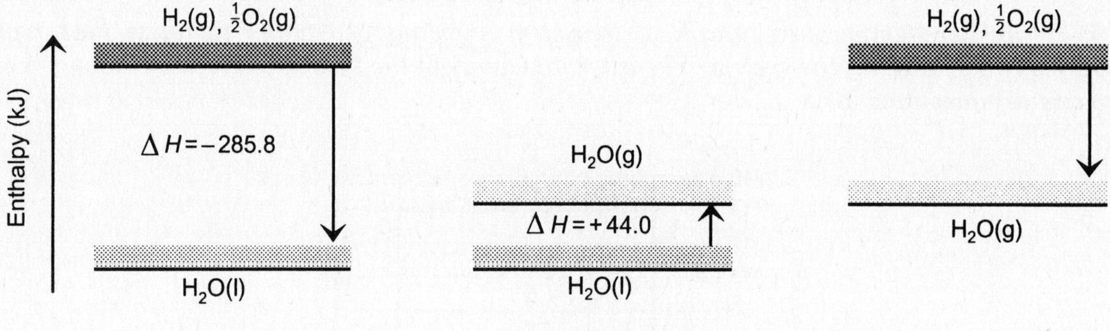

Figure 2

10. Which of the two reactions represents only a physical change?

11. Which reactants have enthalpies of formation of 0 kJ/mol?

12. Which reactions are exothermic?

13. Write the chemical equation for the formation of $H_2O(g)$ from its elements.

14. Determine the enthalpy of formation of $H_2O(g)$ from its elements.

Chapter 27

STUDY GUIDE

27.2 DRIVING CHEMICAL REACTIONS

For each of the following situations determine if the entropy change will be positive (+) or negative (–). Write a plus or minus sign in the blank provided.

____ 1. Calcium chloride crystals are pulverized in a mortar and pestle.

____ 2. Turpentine evaporates.

____ 3. A mixture of iron and sulfur is separated using a magnet.

____ 4. A spark plug ignites a fuel-air mixture in an engine.

____ 5. Turpentine dissolves a paint stain.

____ 6. Melted solder solidifies.

____ 7. Air is heated.

____ 8. Graphite is recrystallized as diamond in a press.

9. Complete the table by naming each state function and giving the appropriate SI unit for a molar quantity.

Symbol	State function	SI unit
ΔH_f°		
ΔG_f°		
ΔS°		
E°		

Answer the following questions.

10. What are standard states of temperature and pressure for measuring thermodynamics quantities?

______

11. What symbol is used to indicate that a quantity has been measured under standard states?

______

12. What is characteristic about the Gibbs free energies of reactions that are spontaneous?

______

13. What is characteristic about the Gibbs free energies of reactions that are at equilibrium?

______

14. What is characteristic about the Gibbs free energies of reactions that are not spontaneous?

______

Place a check mark (✓) in the blank next to each set of conditions that tends to favor spontaneous reactions. Place an X in the blank if the conditions do not favor spontaneous reactions.

		ΔH	T	ΔS			ΔH	T	ΔS
___	15.	<0	low	<0	___	19.	>0	low	<0
___	16.	<0	low	>0	___	20.	>0	low	>0
___	17.	<0	high	<0	___	21.	>0	high	<0
___	18.	<0	high	>0	___	22.	>0	high	>0

23. Use the information in Table 1 to complete Table 2 for the following reaction that takes place under standard measurement conditions.

$$CaCO_3(cr) \rightarrow CaO(cr) + CO_2(g)$$

Table 1

State function	**Compound**		
	$CaCO_3$(cr)	**CaO(cr)**	**CO_2(g)**
$\Delta G_f°$ (kJ/mol)	–1.129	–0.641	–1.650
$S°$ (J/mol·K)	92.8	39.7	885.6

Table 2

Quantity	**Value**	**Quantity**	**Value**
$\Sigma\Delta G_f°{}_{(products)}$		$\Sigma S_f°{}_{(products)}$	
$\Sigma\Delta G_f°{}_{(reactants)}$		$\Sigma S_f°{}_{(reactants)}$	
$\Delta G_r°$		$S_r°$	

24. Is the reaction in question 23 spontaneous? Explain your reasoning.

____

Chapter 28

STUDY GUIDE

28.1 Nuclear Structure and Stability

Answer each question.

1. Name two devices that produce high-energy particles used to bombard nuclei in the investigation of nuclear structure, and compare how the devices are used.

____

____

2. List three things that may happen when a nucleus is bombarded by a high-energy particle, and describe the possible effect of each on the bombarded nucleus.

____

____

____

3. On what three things does the type of change of a bombarded nucleus depend?

____

____

4. State two problems associated with the use of nuclear reactors to produce energy for practical use.

____

____

5. Radioactive element X decays to element Y. A sample initially contains 72 g of element X. At the end of three half-lives, how many grams of element X does the sample contain?

____

Write T *for true or* F *for false. If a statement is false, replace the underlined word or phrase with one that will make the statement true, and write your correction on the blank provided.*

______________ 6. In a synchrotron, <u>drift tubes</u> are used to confine fast moving particles to the ring.

______________ 7. In a <u>circular</u> accelerator, electromagnetic waves are used to accelerate particles to optimum speed.

______________ 8. Neutrons easily penetrate and can be absorbed by nuclei because the neutrons have <u>a negative charge</u>.

______________ 9. <u>Fission</u> involves the breaking apart of a heavy nucleus into two parts of about the same mass.

______________ 10. The emission and absorption of <u>alpha particles</u> are necessary in sustaining a chain reaction.

______________ 11. In nuclear reactions, the law of conservation of <u>mass-energy</u> is upheld.

______________ 12. In a nuclear reactor, a <u>moderator</u> is a device that regulates the rate at which the chain reaction takes place.

______________ 13. The half-life of a radioactive element is the amount of time it takes for half the atoms of a sample to <u>react with other atoms</u>.

Name ______________________ Date ______________ Class ______________

______________ 14. A sample of a radioactive substance has a mass of 10 g. At the end of two half-lives, 0 g of the original sample remains.

______________ 15. Because it does not absorb neutrons, water is often used as a coolant in nuclear reactors.

______________ 16. A synchrotron is a device for controlling fission.

______________ 17. The containment vessel of a nuclear reactor protects people from the heat and radiation of the reactor core.

Match each substance with the part of a nuclear reactor in which the substance is used. Some substances may be used in more than one part of the reactor.

a. moderator
b. control rods
c. coolant
d. fuel
e. containment vessel

___ 18. water

___ 19. uranium-235

___ 20. cadmium

___ 21. plutonium-239

___ 22. molten sodium

___ 23. helium

___ 24. graphite

___ 25. boron

26. Complete the table below to show how atomic number and mass number are affected in various transmutations.

Decay mode	Effect on atomic number	Effect on mass number
α-particle		
	increases by 1	
	decreases by 1	

ame ______ Date ______ Class ______

27. Label the parts of the nuclear reactor shown in the figure and describe the function of each part.

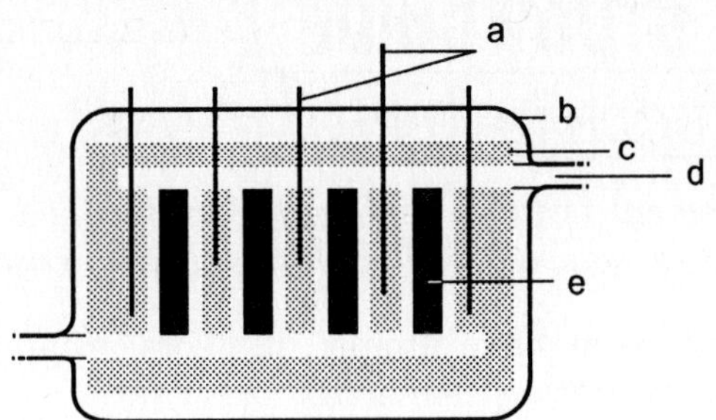

a. ______

b. ______

c. ______

d. ______

e. ______

Use Figure 28.6 in your textbook to answer the following questions about the nuclear decay of uranium-238 to lead-206.

28. What is the total number of alpha particles emitted during the process?

______

29. What is the total number of beta particles emitted?

______

30. What is the atomic number of the nuclide formed at the end of step 5?

______

31. What is the mass number of the nuclide formed at the end of step 8?

______

32. What is the name of the nuclide formed at the end of step 9?

______

33. What is the name of the nuclide with the shortest half-life?

______

34. How is the nuclide produced in step 11 different from the nuclide produced at the end of the decay series?

______

______

Name ______________________ Date ______________ Class ______________

Chapter 28

STUDY GUIDE

28.2 NUCLEAR APPLICATIONS

Write T *for true or* F *for false. If a statement is false, replace the underlined word or phrase with one that will make the statement true, and write your correction on the blank provided.*

______________ 1. Elements that have atomic numbers greater than 92 are known as the transmutation elements.

______________ 2. Because of their relatively large mass and charge, the particles that, once inside the body, are most damaging to cells are beta particles.

______________ 3. Alpha particles are the least penetrating type of radiation.

______________ 4. Radioactive elements are useful as tracers because they can be easily disintegrated.

______________ 5. In fission, the nuclei of small atoms unite to form a larger nucleus.

______________ 6. Fusion generally produces a greater amount of energy per particle than does fission.

______________ 7. Carbon-14 dating is most useful for determining the age of rocks.

______________ 8. Radiation damage to living cells is indicated by the amount of emitted radiation.

______________ 9. Transmutations resulting from "packing" neutrons into the nuclei of certain elements produce elements with atomic numbers lower than those of the original elements.

Answer each question.

10. List three ways in which transuranium elements may be produced.

____

____

11. Differentiate between a gray and a sievert.

____

____

12. List four sources of radiation to which humans are generally exposed.

____

____

____

13. List four ways in which radioactive nuclides are put to practical use.

____

____

14. State three problems associated with the development of plasma fusion reactors.

____

____

____

Chapter 29

STUDY GUIDE

29.1 HYDROCARBONS

Complete each sentence.

1. A chain compound in which all carbon-carbon bonds are single is called an ______ or a ______.

2. Each alkane differs from the next by a(n) ______ group.

3. With increasing molecular mass of compounds within a homologous series, the boiling point ______.

Write the name that corresponds with the following chemical formulas of compounds or radicals.

4. C_5H_{11} ______

5. C_8H_{18} ______

6. C_3H_8 ______

7. C_4H_9 ______

Write T *for true or* F *for false. If a statement is false, replace the underlined word or phrase with one that will make the statement true, and write your correction on the blank provided.*

______ 8. In <u>a branched</u> chain molecule, the numbering of carbon atoms can begin at either end of the chain.

______ 9. When a compound has more than one branch, radicals appear in <u>numerical order</u> in the name of the compound.

______ 10. <u>Prefixes</u> are used in the naming of compounds in which two or more substituent groups are alike.

______ 11. The term *trimethyl* means <u>one three-carbon branch</u>.

______ 12. Isomers <u>are not</u> named according to the total number of carbon atoms in a molecule.

Match the general formulas with the corresponding class of organic compounds.

a. ester
b. ether
c. ketone
d. acid
e. alcohol

___ 13. $R{-}\overset{O}{\overset{\|}{C}}{-}O{-}H$

___ 14. $R{-}\overset{O}{\overset{\|}{C}}{-}O{-}R'$

___ 15. $R{-}O{-}H$

___ 16. $R{-}\overset{O}{\overset{\|}{C}}{-}R'$

___ 17. $R{-}O{-}R'$

Name ____________ Date ____________ Class ____________

Define the following terms.

18. cycloalkanes ____________

19. alkenes ____________

20. unsaturated hydrocarbons ____________

21. alkynes ____________

22. aromatic hydrocarbons ____________

23. How does a benzene ring differ from cyclohexane?

Draw the structural formulas for the following compounds in the spaces provided.

24. benzene

25. cyclohexane

26. phenyl radical

27. methylcyclohexane

28. 1,3-dimethylcyclopentane

29. toluene

30. 2,2,4-trimethylpentane

31. 1-methyl-3-ethylcyclohexane

Answer each of the following.

32. What do benzene, toluene, and xylene have in common?

______

33. What materials is styrene used to make?

______

Name ______________________ Date ______________ Class ______________

Chapter 29

STUDY GUIDE

29.2 OTHER ORGANIC COMPOUNDS

Match the corresponding terms by choosing from the list below.

a. chloromethane
b. methanol
c. 1,2-dichloroethane
d. chloroethene
e. trichloromethane
f. 1,2-ethanediol

___ 1. is also known as chloroform

___ 2. has the formula CH_3Cl

___ 3. is also known as vinyl chloride

___ 4. is also known as ethylene dichloride

___ 5. has the formula CH_3OH

___ 6. is also known as ethylene glycol

Write T *for true or* F *for false. If a statement is false, replace the underlined word or phrase with one that will make the statement true, and write your correction on the blank provided.*

______________ 7. Because aromatic hydroxyl compounds have properties that differ somewhat from those of most alcohols, they are classified as bases.

______________ 8. Ethoxyethane was used for many years as an antiseptic.

______________ 9. Alcohols are neither acidic or basic.

______________ 10. Alcohols with four or more carbon atoms are soluble in water.

Fill in the missing term(s).

11. The most important ketone in industry is ______________.

12. Methanal is more commonly known as ______________.

13. Both aldehydes and ketones are characterized by the presence of a(n) ______________ group.

14. The effect of one functional group on another is called a(n) ______________.

15. Because they do not ionize greatly in water, most organic acids are ______________ acids.

16. The ______________ acid group characterizes most organic acids.

17. Acetic anhydride is made from ______________ by the removal of a molecule of water.

18. In the reaction between ethanol and acetic acid, ______________ and ______________ are produced.

Define the following terms.

19. amine ______

20. primary amine ______

21. amide ______

22. nitrile ______

Write the general formula for each of the following compounds.

23. nitro compounds

24. amines

25. nitriles

Chapter 30

STUDY GUIDE

30.1 ORGANIC REACTIONS

Match the reactants in questions 1-6 with the following products. Then identify the type of organic reaction illustrated in each question.

a.
```
  H H
H-C-C-H + HCl
  |  |
 (O) H
```

b.
```
  H H
H-C-C-OH
  H H
```

c.
```
  H O    H H
H-C-C-O-C-C-H + H2O
  H      H H
```

d.
```
  H H H
H-C-C-C-H
  H I H
```

e.
```
  H H   H
H-C-C=C      + H2O
  H     H
```

f.
```
  H
H-C-H
  |
 (O)   + CH3Cl
```

g.
```
 H Cl H Cl H Cl
-C-C-C-C-C-C-
 H Cl H Cl H Cl
```

______ 1.
```
  H H   H
H-C-C=C     + HI →
  H     H
```

______ 2.
```
         H H      AlCl3
(O) + H-C-C-Cl  ------>
         H H
```

______ 3.
```
  H H  H     H2SO4
H-C-C--C-H  ------>
  H OH H
```

______ 4.
```
H     H         dilute
  C=C    + H2O  H2SO4
H     H        ------>
```

______ 5.
```
  H O        H H
H-C-C-OH + H-C-C-OH →
  H          H H
```

______ 6.
```
H     H     H     H     H     H
  C=C    +    C=C    +    C=C    + . . . →
H     Cl    H     Cl    H     Cl
```

Name ______________________ Date ______________ Class ______________

Write T *for true or* F *for false. If a statement is false, replace the underlined word or phrase with one that will make the statement true, and write your correction on the blank provided.*

______________ 7. In a condensation polymerization reaction, one functional group may be replaced by another.

______________ 8. In an addition reaction, a double bond between carbons is created.

______________ 9. A reaction in which water is removed from an alcohol is an esterification reaction.

______________ 10. Soap is made by the polymerization of fatty acids.

______________ 11. Like molecules called *monomers* chain together to form polymers.

______________ 12. When an ester is deformed by an outside force, it returns to its original shape when the force is removed.

______________ 13. Rayon is made from reconstituted cellulose.

______________ 14. Many compounds containing benzene readily take part in addition reactions.

______________ 15. In the fractional distillation of petroleum, the boiling point of the gasoline fractions must be lower than the boiling point of the fuel oil fractions.

______________ 16. Cracking is a process used to increase the yield of natural rubber.

______________ 17. A thermoplastic material hardens upon heating.

______________ 18. The higher the octane rating, the greater the percent of the fuel that burns evenly.

Chapter 30

STUDY GUIDE

30.2 BIOCHEMISTRY

Write T for true or F for false. If a statement is false, replace the underlined word or phrase with one that will make the statement true, and write your correction on the blank provided.

______ 1. Enzymes are lipid-based biological catalysts.

______ 2. Proteins are composed of nucleic acids linked by peptide bonds.

______ 3. Steroids, some vitamins, and fats are classified as carbohydrates.

______ 4. Carbohydrates are more soluble in water than in nonpolar solvents.

______ 5. Protein synthesis in the human body is controlled by amino acids.

______ 6. Polysaccharides are formed by addition polymerization.

______ 7. Cartilage and tendons are composed mainly of cellulose.

Answer the following questions.

8. State three ways in which proteins may differ from one another.

______

______

______

9. List three glucose-based polysaccharides found in living things.

______

10. State three problems associated with the use of biomaterials.

______

______

______

11. List four uses of biomaterials.

______

______

12. Compare the chemical composition of RNA with that of DNA.

______

______

______

______

Name ______________________ Date ______________ Class ______________

Refer to Table 30.1 in your textbook to answer the following questions.

13. Draw a structural formula for the amino acid serine.

14. Draw a structural formula for the amino acid glycine.

15. Use structural formulas to show how serine and glycine may combine to form two different dipeptides.